Hazard Analysis Primer

Clifton A. Ericson II

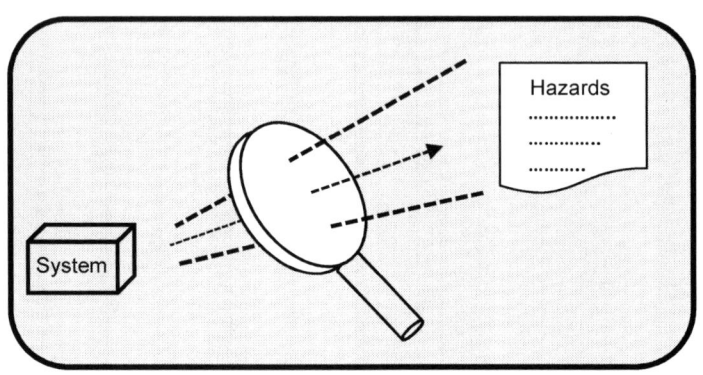

Copyright © 2012 Clifton A. Ericson II

All rights reserved.

ISBN-10: 1470092530

ISBN-13: 978-1470092535

Printed By CreateSpace Inc., Charleston, NC

CreateSpace.com

CONTENTS

1	The Safety Equation	1
2	Hazard Ambiguity	7
3	Systems Engineering Link	17
4	Hazard Theory	31
5	Hazard Risk Theory	45
6	System Mishap Model	63
7	Hazard Analysis Theory	69
8	Hazard Analysis Steps	87
9	Hazard Analysis Tools	91
10	Hazard Analysis Techniques	97
11	Hazard Recognition Checklists	109
12	Common Hazard Analysis Mistakes	117
13	Hazard Causal Factors	121
14	Failure Data	135
15	Hazard Analysis Questions	141
16	Hazard Analysis Examples	163
17	Mishap Investigation Models	173
18	Summary	175
A	Mind Mapping and HA	179
B	References	183
C	About the Author	185
D	Index	187

ACKNOWLEDGEMENTS

I would like to thank my good friend Adam Gambriell who graciously edited my draft manuscript and provided meaningful comments and suggestions.

PREFACE

Accidents are not acts of God; they are acts of carelessness – that is, carelessness in system design. These are designs that allow accidents to happen when components fail, environments are adverse or humans perform critical errors. These acts can be recognized as hazards within a system design, and they can be eliminated or controlled through the hazard analysis process.

Hazards, mishaps and risks are all too common and they seem to have become a complacently acceptable way of life. Many hazards do not have to be allowed to persist and can be eliminated. On the other hand, there are certain hazards that cannot be eliminated because of unique system factors involved. Although these types of hazards must be tolerated, the good news is that the risk they present can be controlled and reduced to an acceptable level (i.e., small likelihood of occurrence). Knowing the difference between these types of hazards, and how to identify, eliminate or mitigate them is an important aspect of safety that requires hazard analysis.

Hazards are a result of the way in which systems are designed (i.e., the system architecture). We utilize systems on a daily basis, and systems expose us to many different types of hazards and potential mishaps. Most systems involve hazardous components and safety-related functions, which means that these systems will naturally have inherent hazards as part of their design. These hazards must be eliminated or controlled if mishaps are to be prevented, which requires hazard identification via hazard analysis.

Since hazards and risk are a result of system design, they can also be changed through system design. System safety is an engineering discipline that is applied during the design and development of a product or system to proactively identify and eliminate/mitigate hazards, and thereby eliminate/reduce the risk of potential mishaps and accidents. Reacting after a mishap is contrary to the principles of system safety. System safety is ultimately about saving lives. Hazard identification and control are key elements in the system safety process.

Hazard identification is achieved through the hazard analysis process. The practice of hazard analysis requires detailed knowledge and understanding of what comprises a hazard. It also requires an understanding of the tools, techniques and processes involved in the hazard analysis process. Hazard analysis is not an entirely simple process. There are many complexities involved, as well as misunderstandings regarding the hazard concept. All of these issues are addressed in this book.

Hazard analysis is primarily about two things: 1) finding the hazards in a system and 2) eliminating or reducing the risk presented by these hazards. In order to perform a credible hazard analysis an analyst must be skilled in three things: 1) systems analysis, 2) hazard theory and 3) risk theory. This means an analyst must be able to grasp and understand the design of a system; how it works and how it is operated. An analyst must also understand what a hazard really is and how to identify and describe one in a realistic manner. Finally, the analyst must be able to calculate the risk presented by a hazard and then loop back to the system design to know how to reduce the risk. This book describes all of these aspects in detail.

This book is for engineers, analysts and managers who are confronted with the responsibility of developing safe systems and products through the process of hazard analysis. This book describes the hazard concept and hazard analysis process from A to Z.

Clif Ericson
Fredericksburg, VA
February 7, 2012

CHAPTER 1

THE SAFETY EQUATION

1.1 Introduction

Lives, property, resources and money are lost every day to mishaps and accidents[1]. The premise of system safety engineering is that mishaps are preventable. This is based on the fact that hazards are a prerequisite to a mishap, and that hazards can be identified and controlled before they become a mishap. In order to eliminate or mitigate hazards they must first be identified; hazard identification is achieved through the hazard analysis (HA) process.

HA involves looking into the future of a systems operation, based on the system design, and identifying potential mishaps that can result from both normal and abnormal conditions. These conditions can result from failures, misuse, erroneous use, wearout, design errors, etc. Once a credible hazard is identified the next step is to design-out the hazard or design to guard and protect against the hazard. Hazard identification, evaluation and control form the backbone of the system safety process.

Although simple in theory, HA is a complex process, fraught with misunderstandings and incorrect approaches. Performing a good HA is not a simple or trivial task; it requires foresight, planning, organization and a total systems viewpoint in order to achieve uniformity, consistency and full system coverage. Quite often safety analysts jump too quickly into a HA and immediately start identifying what they think are hazards, without first considering or developing the system mishap models needed to organize hazards. This often leads to incorrect hazards, hazard overlap, hazard gaps and hazard-risk mismatches.

There is no HA shortcut; HA requires knowledge, skill, experience, commitment and effort. This chapter focuses on the safety equation, which involves the interrelationship of systems, hazards, mishaps and risk. The

---

[1] In system safety, the terms *accident* and *mishap* are considered synonymous; the term mishap will be used herein to represent both.

safety equation establishes the fundamental purpose for HA. Many of the concepts and concerns presented in this chapter will be expounded upon in later chapters.

1.2 Hazards, Mishaps, Risk and Safety

We live in a complex and imperfect world, filled with dangers (i.e., hazards) that threaten our lives. Hazards are unsafe design conditions that can ultimately result in mishaps if left unchecked. Mishaps are undesired events responsible for the death, injury, damage and loss of property, lawsuits and many other tangible and intangible losses. On a personal level hazards threaten our lives, health, families and finances. On a larger scale hazards endanger corporations, localities and countries. Hazards, mishaps and risk may be a way of life, but that does not mean they cannot be intentionally controlled and constrained.

Mishaps are not just random uncontrollable events, or events due solely to human error or human carelessness, or events resulting from extreme environmental conditions, such as tornadoes and hurricanes. Mishaps are not acts of God or pre-ordained destiny; they are the result of man-made systems containing hazard-related flaws. Mishaps are *predesigned* undesired events that have occurred. Mishaps are predictable, and hazards are the blueprints for predicting potential mishaps. When we develop the systems that we use daily, we inadvertently design-in hazards into these systems. These hazards, if not properly dealt with eventually result in mishaps. The likelihood and severity of a mishap is based upon the risk presented by the hazard forming it. As a blueprint for a potential mishap, a hazard contains the specific causal factors, conditions and outcome for a predicted mishap.

Safety involves a unique relationship between hazards, mishaps and risk. Typically, hazards stem from the systems we design, develop and build. Systems are designed to be safe or unsafe; if systems are not intentionally designed to be safe, they are by default potentially unsafe. System safety engineering has a defined and rigorous process for intentionally designing safe systems and HA is a key element of that process.

In other words, preventing mishaps requires systems HA for the identification and elimination/mitigation of hazards. And, HA utilizes the interrelated concepts of: hazard, mishap and risk. Safety must be *earned* through the rigorous application of the system safety process involving detailed HA. Safety is not free, but it is much cheaper than the cost of mishaps.

1.3 The Safety-Risk Equation

The safety equation is a characterization of the basic components of safety: hazards, mishaps, risk and system design. It is actually system design that creates and shapes the hazards and their associated levels of risk. The following are the fundamental equations of safety:

Safety = Function (System Design, System Hazards, Hazard Risk)
Hazard = Function (System Design, Hazard Components, Hazard Outcome)
Mishap = Function (Hazard, Risk)
Risk = Function (Hazard Likelihood, Hazard Severity)
Hazard Components = Function (HS, IM, TTO)   [explained in chapter 4]

The idea behind these equations is that safety is really based upon system design and the hazards created within the design, i.e., safety is a function of design, hazards existing within the design and the amount of risk presented by the hazards. Hazards are the precursor to mishaps and risk is a measure of the criticality of a hazard and its concomitant potential mishap. Mishaps do not have to be tolerated; they can be prevented by identifying the hazards in the system design, and then modifying the system design to eliminate or mitigate the hazard and its associated risk. The hazard components are the set of unique conditions creating and forming the hazard, which are determined by the system design. Risk is the estimated likelihood of a hazard becoming a mishap, combined with the severity the mishap consequences should it occur.

The purpose of safety is to save lives by preventing mishaps. Mishaps can be prevented; they do not have to be accepted. Mishaps and hazards are directly linked. HA is a strategic element in the system safety process, and the design of a system is key to both safety and HA.

Safety is quite often seen as being somewhat binary, something is either safe or else it's unsafe. This is thinking in absolutes. Safety is not that simple, it's more complex; things cannot usually be made absolutely safe. Safety is really a degree of risk; the smaller the risk the safer something is, the larger the risk the more unsafe it becomes. System safety engineering understands that risk (and thus safety) is a quantifiable and measurable quantity, which can be engineered into a product or system.

Figure 1.1 demonstrates that thinking of safety in terms of safe and unsafe results in an undefinable measure of safety. On the other hand, thinking of safety in terms of risk provides a definable and workable measure of safety. Everything lies somewhere on the safety-risk continuum. The object of HA is to determine that position and change it if necessary.

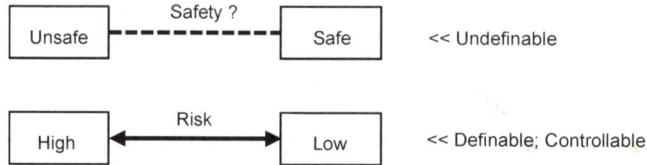

Figure 1.1 – Risk as a Measure of Safety

1.4 What is Hazard Analysis?

Hazard identification and control are key elements in the system safety process. System safety is a formal and systematic engineering and management process for proactively *making* a system safe. The bottom line is that hazards (and mishaps) can be eliminated, or their risk reduced, when the proper system safety processes are conscientiously applied.

HA is the act of performing a special analysis for the identification and evaluation of hazards. A hazard analysis typically tries to answer the following questions:

1) What can go wrong that will lead to a mishap
2) How exactly will it happen
3) What are the consequences
4) What is the risk involved
5) Is the risk acceptable
6) How can the design be modified to eliminate or reduce the risk

The primary purpose of HA is to identify hazards and to obtain sufficient hazard data for risk assessments. HA is applied to hardware, software, functions, procedures and human tasks. HA can be applied at all stages of the system lifecycle; the HA process becoming more detailed and accurate as the design progresses and more detailed system design information becomes available. Different HA techniques and approaches to hazard identification may be required at different stages of the system life cycle to ensure all types of hazards are identified. HA involves the following basic elements:

- Acquiring system data, knowledge and understanding.
- Identifying the hazards that exist within the system design.
- Determining the hazard causal factors and chain of events involved.

- Determining the likelihood [probability] of the hazard occurring.
- Determining the potential consequences resulting from an occurrence of the hazard.
- Investigating any safeguards already in place to address the hazard.
- Assessing the risk presented by the identified hazard and determining if it is acceptable or not.
- Establishing design requirements to mitigate the hazard (new safeguards) and its associated risk.

In order to design-in safety, hazards must be designed-out (eliminated) or mitigated (reduced in risk), which can only be accomplished through HA. Hazard identification is a critical system safety function and is one of the basic required elements of a System Safety Program (SSP). HA assesses the functional and safety characteristics of the system or product.

HA provides the basic foundation for system safety. HA is the systematic examination of a system, performed to identify hazards, hazard effects and hazard causal factors. HA is used to determine system risk, to determine the significance of hazards, and to establish design measures that will eliminate or mitigate the identified hazards. HA is used to systematically evaluate systems, subsystems, facilities, components, software, personnel, and their interrelationships, with consideration given to logistics, training, maintenance, test, modification, and operational environments. In order to effectively perform HAs it is necessary to understand what comprises a hazard, how to recognize a hazard, and how to define a hazard. To develop the skills needed to identify hazards and hazard causal factors it is necessary to understand the nature of hazards as they exist within a system design.

1.5 Hazard Analysis Complexities

There are many challenges related to HA. Many of these challenges involve issues revolving around the interpretation of a hazard; however, there are also many methodology challenges in addition. Example HA challenges and issues include:

- Hazard definition ambiguity; what are the specific characteristics and requirements of a hazard?
- Hazard description; should a hazard description be terse and ambiguous or long enough to be meaningful and well understood?

- Do multiple causal factors support a single hazard or multiple similar hazards?
- Hazard count: is a large quantity of hazards a bad thing; should hazards be combined to make the system design look better?
- Hazard risk roll-up or roll-down; at what system level should risk be tabulated and evaluated?
- Risk summing; should hazard risks be summed into one risk value; is it possible or even meaningful?
- HA Types and Techniques; what do they mean and how should they be applied?
- HA methodologies: is one HA technique adequate or should more than one be performed?
- Hazard recognition; how are hazard identified?
- Top Level Mishaps (TLMs); what are TLMs, how are they used and how do they relate to hazards?

1.6 Summary

Hazards are emergent properties of a system, which we create in the design; these properties should be intentional rather than unintentional. Hazards are predictable, and what can be predicted can also be eliminated or controlled. HA is a tool for hazard identification, assessment and elimination or control. Most systems have hazards due to complexity and the use of hazardous energy sources.

In order to perform a credible, thorough and valid HA the following considerations are necessary:

- A grounded understanding of the definition of a hazard
- The knowledge of what specifically comprises a hazard
- The skill to correctly describe (write) a hazard
- The skill to develop a system hazard-mishap model
- The proficiency for hazard recognition
- The skill to develop hazards at the correct system level

CHAPTER 2

HAZARD AMBIGUITY

2.1 Introduction

Everyone knows what a hazards is, right? It's something that is dangerous or presents a danger. The problem is it's not that simple. A hazard is somewhat of a complex entity, as well as a complex concept, that requires a clear and precise definition. It seems in reality that many people do not fully understand the concept of a hazard, thereby making the task of HA much more difficult. In order to understand the risk presented by a hazard, as well how to eliminate or mitigate a hazard, a certain amount of information is required of a hazard. This chapter focuses on the problems pertaining to the definition of a hazard, which ultimately has a significant impact on hazard identification and hazard risk management.

2.2 Hazard Definition Confusion

The confusion over what constitutes a hazard has been a major drawback to meaningful HA for some time. In order to identify hazards the concept of a hazard must be clearly and precisely understood. There are many factors causing confusion, however, one of the biggest confusion factors is the definition of a hazard.

Presently there is misunderstanding and confusion regarding the definition of a hazard, with a wide spectrum of definitions with significant differences. There is a profusion of definitions available, yet there is little common agreement or consensus among these definitions. Current definitions do not adequately fit the need for hazard identification, especially for a scientific community trying to conscientiously identify and eliminate hazards.

For example, consider the following definitions that currently exist within industry and society:

Hazard: "A source of danger" (1).

Hazard: "Any real or potential condition that can cause injury, illness, or death to personnel; damage to or loss of a system, equipment or property; or damage to the environment" (2).

Hazard: "A condition resulting from failures, external events, errors, or combinations thereof where safety is affected" (3).

Hazard: "The presence of a potential risk situation caused by on unsafe act or condition" (4).

Hazard: "The presence of a potential risk situation caused by an unsafe act or condition. A condition or changing set of circumstances that presents a potential for adverse or harmful consequences; or the inherent characteristics of any activity, condition or circumstance which can produce adverse or harmful consequences" (5).

Software Hazard: "A software condition that is a prerequisite to an accident" (6).

Hazard: "A source of danger (i.e., material, energy source, or operation) with the potential to cause illness, injury, or death to personnel or damage to an operation or to the environment (without regard for the likelihood or credibility of accident scenarios or consequence mitigation)" (7).

Hazard: "A hazard is a condition or changing set of circumstances which present an injury potential" (8).

Hazard: "A hazard is a situation that poses a level of threat to life, health, property, or environment. Most hazards are dormant or potential, with only a theoretical risk of harm; however, once a hazard becomes "active", it can create an emergency situation. A hazard does not exist when it is not happening. A hazardous situation that has come to pass is called an incident. Hazard and vulnerability interact together to create risk" (9).

Hazard: "The potential for harm; also see Contributory Hazard, Primary Hazard. A hazard is not an accident. Per FAA Order 8040.4 a "condition, event, or circumstance that could lead to or contribute to an unplanned or undesired event." Anything, real or potential, that could make possible, or contribute to making possible, an accident. A condition that is prerequisite to an accident" (10).

Primary Hazard: "A primary hazard is one that can directly and immediately results in: loss, consequence, adverse outcome, damage, fatality, system loss, degradation, loss of function, injury, etc. The primary hazard is also referred to as: catastrophe, catastrophic event, critical event, marginal event, and negligible event" (10).

Contributory Hazard: "The potential for harm. An unsafe act and/or unsafe condition which contributes to the accident, (see cause, root cause, contributory events, initiator; the potential for adverse energy flow to result in an accident.) A hazard is not an accident. A failure or a malfunction can result in an unsafe condition, and/or unsafe act. Human error can result in an unsafe act. Contributory Hazards define the contributory events that lead to the final outcome. For simplicity, Contributory Hazards can also include Initiating Events and Primary Hazards. Sequential logic defining the Hazardous Event should remain consistent throughout the hazard analysis process" (10).

The reference sources for the above definitions are taken from the following reputable industry sources:

1) Merriam-Webster's Desk Dictionary, 1995.
2) DoD MIL-STD-882D, *Standard Practice For System Safety*, Feb 2000.
3) SAE ARP-4754A, SAE Aerospace Recommended Practice, *Guidelines for Development of Civil Aircraft and Systems*, December 2010.
4) DHB-S-001, *NASA System Safety Handbook*, NASA Dryden, March 2, 1999.
5) NASA-GB-8719.13, *NASA Software Safety Guidebook*, March 31, 2004.
6) IEEE Std 1228-1994, *IEEE Standard for Software Safety Plans*, August 9, 1994.
7) DOE-STD-3009-94, *Preparation Guide For U.S Department of Energy Nonreactor Nuclear Facility Documented Safety Analyses*, April 2002.

8) H. M. Philo, et al, *Lawyers Desk Reference*, 9th edition, 2001, volume 2, chapter 5, page 3.
9) Wikipedia, http://en.wikipedia.org/wiki/Hazard, obtained on November 22, 2011.
10) Federal Aviation Administration (FAA), *System Safety Handbook: Practices and Guidelines for Conducting System Safety Engineering and Management*, Appendix A – Glossary, Dec. 30, 2000.

It can be seen from these definitions that they all seem to allude to a common concern for safety, yet they are not in common agreement and they are not clear and precise. Some are contradictory, some are misleading and some are patently wrong. For example, the Wikipedia definition states that "A hazard does not exist when it is not happening"; not only is this ambiguous and misleading, but it is also incorrect. A hazard either exists or it doesn't. Another definition states that a hazard can be a "changing set of circumstances". This definition is unclear and misleading; the only circumstances that change are when the hazard components move from a dormant state to an active mishap state, but there are other factors involved in conjunction with the changing circumstances. I cite these definitions to illustrate that we all tend to fundamentally understand the concept of the term hazard, yet we cannot appropriately define it and we are easily mislead by industry and social definitions.

These definitions open a host of questions. For example, using these definitions, how would an analyst accurately identify hazards in a precise and consistent manner? What exactly is the *condition* in a hazard; what is a hazard comprised of and how is it characterized? Why aren't all hazard definitions identical? How does one deal with primary and contributory hazards as per the FAA definition? Are the current definitions really meaningful and useful?

2.3 Hazard Confusion Examples

The lack of a firm technical definition of a hazard, and the resulting hazard confusion, has led many analysts to incorrectly identify hazards. Over the years I have reviewed many different HAs, performed by different individuals and groups on a diversity of different systems. To help demonstrate the confusion over what comprises a hazard, the following list is provided of hazards taken from real-life HAs:

- Gasoline
- High voltage

- Broken glass
- Signal MG71 occurs
- Oil spill
- Brake failure
- Cracked bolt
- Inadvertent missile ignition
- Controlled flight into terrain
- Explosive uncontainment of gas charge used for emergency gear extension purposes
- Explosive decompression hazard servicing tires
- Broken ladder rung
- Safe and arm device fails
- Electrocution

There is a common thread through these example postulated hazards; they all allude to something that seems unsafe and possibly presents a danger. Yet, the real danger and threat are not entirely clear. As will be seen later in this book, these are really pieces of hazards and mishap categories. These postulated hazards, produced by experienced and inexperienced analysts, demonstrates the confusion created by poor hazard definitions.

One very easy way to know if a hazard has been identified is to determine if the hazard description provides enough information to understand the specific danger and calculate the risk presented. All hazards contain potential mishap risk, and this information can only be obtained from the hazard description. For example, if I postulate "high voltage" as a hazard, then what is the specific danger and risk presented, and how is the hazard best eliminated? There is obviously not enough information provided in this example to determine the risk presented by the hazard or to eliminate the hazard causal factors. Unfortunately this type of example happens all too often.

Care must be taken to differentiate between *pseudo hazards* and true hazards. For example, consider the examples shown in Table 2.1:

Table 2.1 – Pseudo Hazards vs. True Hazards

Pseudo Hazard	True Hazard
Electrocution	Worker makes contact with exposed high voltage conductor.
Hearing damage	Worker exposed to high noise environment without hearing protection.
Fall injury	Worker trips and falls from platform because no barriers or harness protection are present.

Pseudo hazard descriptions typically address only the mishap outcome, whereas a true hazard description describes all of the relevant factors involved.

2.4 Hazard Definition Issues

In general HA seems like a relatively simple concept – analyze a system using a rigorous methodology and identify all hazards within, and related to, the system. There are, however, some basic issues that must be addressed before delving into the HA process. These issues stem from the varying array of definitions for a hazard. Because hazards are misunderstood, several different hazard models have been developed. Many of these models are either incorrect or inadequate. The following chapters will develop a recommended definition of a hazard based on hazard and risk theory, which helps the HA process significantly.

Hazards are a key concept in system safety and mishap prevention. A mishap cannot occur without the pre-existence of a hazard. Fully understanding the concept of a hazard, or what comprises a hazard is key to effective HA. There are many different issues regarding hazard understanding, each of which will eventually be addressed in this book. These issues include the following:

- Too many diverse and inconsistent hazard definitions exist

- Existing hazard definitions do not adequately describe a hazard, particularly for engineering purposes

- The vagueness of current hazards definitions leaves hazard identification open to interpretation and misunderstanding

- Many accident models and hazard models exist, each of which do not adequately define the concept of a hazard

- Many different hazard categories have been proposed, such as electrical, biological, food, energy, etc. which tend to cloud hazard understanding

- Most definitions fail to show an understanding of the relationship between hazards, mishaps and risk

What is needed to aid the HA process is a clear and concise hazard definition that is universally agreed upon. This definition must provide a technical and engineering baseline for a hazard, which includes the hazard causal factors, the potential hazard outcome and the potential risk involved. There must be a clear understanding of what technically comprises a valid hazard in order to identify and control hazards.

2.5 Hazard Model Confusion Example

A major hazard confusion factor involves the various perceptions of what comprises a hazard. Figure 2.1 contains an example hazard diagram for the hazard of "Fuel Tank Rupture", which is documented in a reputable industry source[2]. This diagram shows an Initiating Hazard, Contributory Hazards and Primary Hazards, all for a single hazard scenario. This concept causes considerable confusion, because a hazard cannot contain sub-hazards. This a very good diagram of the ingredients involved in the hazard, but the overall concept is causing a problem of clarity. Rather than being contributory or sub-hazards, the noted items are really *events* forming the specific conditions causing the *single* hazard. What are referred to as initiating hazard, contributory hazards and primary hazards are really the causal factors and outcome for a _single_ hazard.

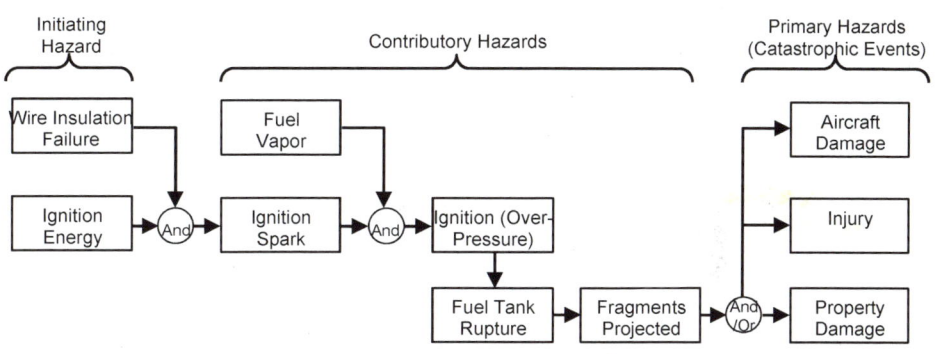

Figure 2.1 – Multiple Hazard Model Concept (source: FAA)

Figure 2.2 shows the same diagram, but with the hazard components identified in the more recent HS-IM-TTO hazard model, which is fully explained in chapter 4. This model identifies the hazard components, which include hazard source (HS), initiating mechanisms (IMs) and target/threat outcome (TTO).

---

[2] *System Safety Handbook: Practices and Guidelines for Conducting System Safety Engineering and Management*, FAA, Chapter 7, Figure 7-2, Dec. 30, 2000.

Figure 2.2 converts the multiple hazard concept of Figure 2.1 into a more accurate single model hazard that can be more easily understood and rationalized.

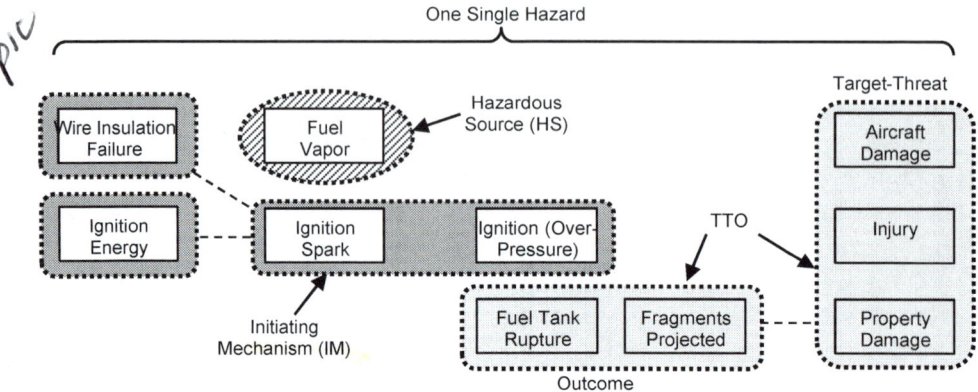

Figure 2.2 – Single Hazard Model Concept

The HS-IM-TTO hazard model is more logical; it breaks a hazard into three required components, each of which is necessary for a hazard to exist. It ties together contributing events, factors, conditions and potential outcomes. The IM involves the changing circumstances that transform the hazard from a dormant state to an active mishap state.

The hazard shown in Figure 2.2 can be redrawn for clarity as shown in Figure 2.3.

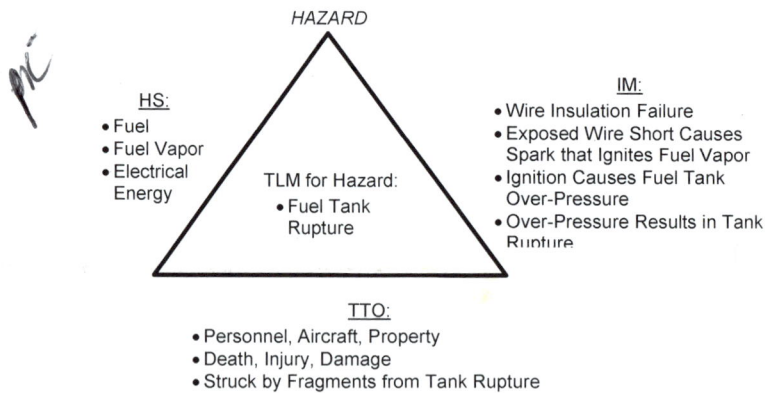

Figure 2.3 – Hazard Triangle for HS-IM-TTO Model

Note that it is not always clear whether an item should be a HS or IM. In Figure 2.2 "ignition energy" is shown as an IM, however, it could also be shown as a HS. The important point is that it is identified in the hazard description.

2.6 Mishap Confusion

Another possible reason hazards are misunderstood is because, like hazards, mishaps are also largely misunderstood. There is much confusion over what comprises a mishap and how to identify the root-cause factors of a mishap following the mishap occurrence. See chapter 17 for a discussion on several *a posteriori* mishap models.

2.7 Summary

Some hazard definitions state that a hazard is "any *real* or *potential* condition that can cause ..." An interesting fact is that a hazard is both real and potential, not one or the other. If a hazard exists, it is real; and, all hazards are a potential for a mishap. Another fact to remember is that a hazard either exists or it does not exist, there is no in between. Whether or not a hazard becomes a mishap depends upon the likelihood of the hazard moving from the dormant state to the active state.

As previously shown, the various hazard definitions in use are misleading, uninformative and contribute to the confusion regarding hazard understanding. The key question that must be answered before hazard identification and elimination can be successful is: what's in a hazard? What defines a hazard and what are the constituent elements of a hazard? It is clear that hazards are the precursor to mishaps; a mishap cannot occur unless a hazard first exists. A hazard is a blueprint of a potential mishap. Chapter 4 delves into how to establish and read the hazard blueprint.

Figure 2.4 demonstrates the difference between a hazard and a mishap. In this example the hazard contains all of the conditions necessary for a mishap: high voltage lines, a metal ladder extension truck and an employee with instructions to repair the power line. The mishap results when the ladder is extended too far and makes contact with the high voltage lines, thereby conducting electricity to the employee. The hazard can be foreseen through proper HA and it can be mitigated by through design safety features, such as a non-conducting bucket and a power line proximity warning device.

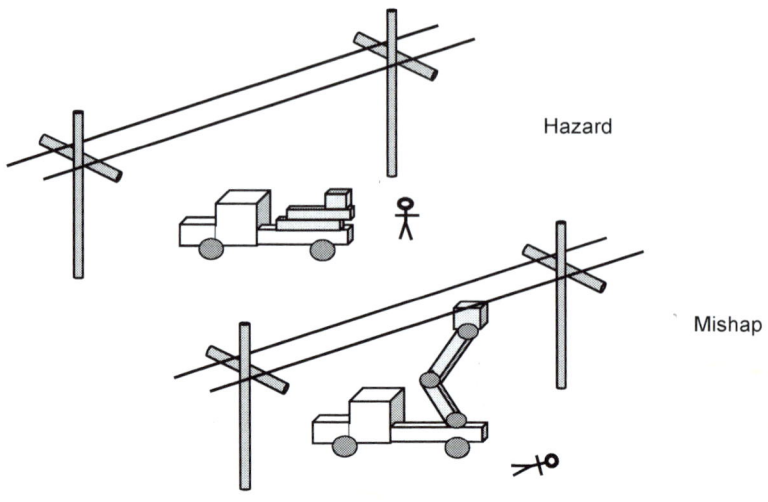

Figure 2.4 – Hazard/Mishap Example

CHAPTER 3

SYSTEMS ENGINEERING LINK

3.1 Introduction

Systems are the fabric of life; our lives revolve around systems. Unfortunately, systems also expose us to many different types of hazards. Because we continually use and interface with systems and systems-of-systems, it is critical to develop systems that are safe. Not only is system design knowledge essential to system safety and hazard prevention, but understanding the systems engineering process is also critical. HA is aided through the use of the many products produced by the systems engineering process. System safety and systems engineering are partners in system development. This chapter focuses on the systems engineering link with system safety engineering and HA.

3.2 Systems and Safety

Systems contain the factors that spawn hazards, which ultimately result in undesired events and mishaps if not countered. Understanding all of the factors involved with a particular system design is necessary for the hazard identification and risk mitigation process. This means thoroughly understanding system design, attributes and complexity is necessary for the HA process. In order to effectively eliminate hazards the total system must be addressed (system safety tenet). It is not effective to pick one small part and try to fix it, in isolation, in order to eliminate a hazard; there are too many system interrelationships involved.

System safety is the engineering discipline applied during the design and development of a product to develop safe systems. Systems can only be made safe when the hazards existing within the system design are identified and mitigated or eliminated. Systems contain the ingredients and factors that produce hazards and mishaps. Systems can fail, malfunction and/or be erroneously operated. Systems utilize components that are inherently hazardous. Systems are often designed with built-in safety flaws. To design systems that work correctly and safely, a safety analyst needs to understand how things can go wrong in a system and how to correct them; through the system HA process.

Systems engineering is the discipline applied to design and develop a product or system; it involves the tasks and tools utilized to cost-effectively build a system. Many of products and tools used in the systems engineering process can be used in the HA process, thus their understanding is imperative.

3.3 System Definition

A system is an integrated composite of components that provides function and capability to satisfy a stated need or objective. A system is a holistic unit that is greater than the sum of its parts. Systems have structure, function, behavior, characteristics and interconnectivity. They vary in size, purpose, type, complexity and safety criticality. Modern day systems typically comprise people, products, processes and environments that together provide the desired and intended capability.

Figure 3.1 characterizes the general aspects and components of a system. The key components comprising a system are hardware, software and people. A system has purpose and structure, and the components and arcitecture establish certain qualities and characteristics unique to the system design. System objectives are achieved via the operations, procedures and functions designed-into the system. The system has its own internal environment, and is impinged upon by external environments, systems and people. A very important safety aspect of a system involves the particular hazardous assets within the system.

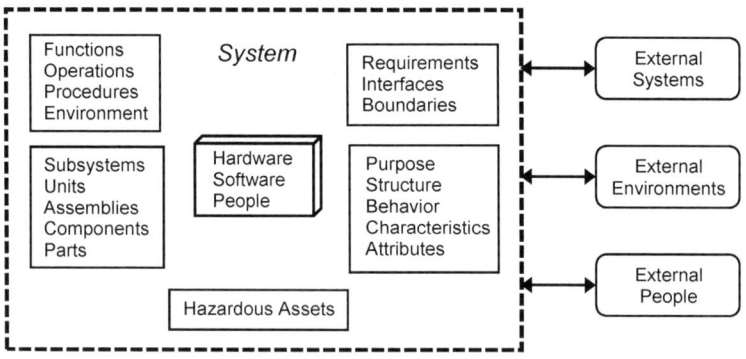

Figure 3.1 – System Characterization

In short, a system is any group of interrelated, interacting and interdependent parts that form a complex and unified whole that has a specific function or purpose that is greater than that of the individual parts. If the parts are not interacting and interdependent, it is not truly a system; it is merely a collection of parts.

Understanding the elements of a system is necessary in the performance of HA. A system typically comprises any combination of the following elements, each of which must be clearly understood:

- Hardware (electrical, hydraulic, structures)
- Software (program, segment, unit, module, logic, algorithms)
- People (operators, testers, maintainers, manufacturers)
- Functions (operations, modes, phases, tasks)
- Hazardous Assets (explosives, fuel, electricity)
- Processes (course of action, timing, material combining)
- Procedures (instructions, tasks, manuals, warning notes)
- Interfaces (hardware, software, documentation, communications)
- Facilities (building, location, storage, transportation)
- Boundaries (physical, theoretical, limitations)
- Environment (weather, temperature, vibration)
- Hierarchy (subsystems, units, assemblies, components)

Two distinguishing characteristic of systems are that a) the whole is more than the sum of the parts, and b) what is best for the parts is not necessarily best for the overall system. As might be expected, these two system characteristics are important in safety analysis and in identification of hazards. System safety works best at the system level because, as has been long recognized, many hazards involve unique system interrelationships between the parts, as opposed to the parts in isolation. Although system safety must focus on the system, it must also evaluate the parts as well as the system as a whole.

3.4 Hazardous System Assets

In order to achieve their desired objectives, systems are often obliged to utilize hazardous assets in the system design. Hazardous assets are the things that present the basic danger source in a hazard. These hazardous assets comprise hazardous components, materials, operations and functions. Example hazardous components include items such as gasoline, nuclear material, high voltage, high pressure, etc.

Hazardous assets bring with them the potential for many different types of hazards, which if not properly controlled can result in mishaps. In one sense, system safety is a specialized trade-off between utility value and harm value, where utility value refers to the benefit gained from using a hazard source and harm value refers to the amount of harm or mishaps that can potentially occur from using the hazard source. For example, the explosives in a missile provide a utility value of destroying an intended enemy target; however, the same explosives also provide a harm value in the associated risk of inadvertent initiation of the explosives and the harm that would result. The system safety process involves balancing utility and harm values, through HA and design safety defensive mechanisms.

Systems using hazardous assets will naturally have inherent hazards as part of their design. Although these are residual hazards that cannot be eliminated, they can be controlled to prevent mishaps from resulting. HA is an essential component in identifying and controlling these hazards.

Hazard asset categories include, but are not necessarily limited to, the following:

- Energy sources
- Safety-critical functions
- Safety-critical operations
- Hazardous materials
- Hazardous equipment
- Biological sources

3.5 System Views

Chapanis[3] recognized that systems can be perceived from three different perspectives, each of which provides a different viewpoint and understanding of the system. These system viewpoints include:

1) **Physical** – the architectural view that depicts what the system contains and how it is constructed.
2) **Functional** – describes what the system must do in order to produce the required system behavior, broken down into functions with input, output and transformation rules.
3) **Operational** – defines how the user will interface with, and operate the system, including instructions, conditions, parameters and limitations.

---

[3] Alphonse Chapanis, *Human Factors in Systems Engineering*, New York, Wiley, 1996.

It is important that HA consider and evaluate a system from each of these three perspectives in order to ensure complete safety coverage of the system. This is why HA must consider and focus on system functions, system operations, and system components, including hazardous energy sources. This also explains why HA requires more than one type of HA in order to identify all hazards; one HA type alone does not typically provide sufficient hazard identification coverage for all the many facets of a system.

There are some additional systems perspectives that must be considered in HA, such as:

- Software – this view looks at the system software that essentially controls computer controlled systems.
- Environment – this view looks at the various environments that the system will encounter (internal and external).
- Human – this view looks more closely at human performance in the system and the effect of potential human errors.
- Organizational – this view considers the organizational and management aspects affecting a hazard.

3.6 System Lifecycle

All systems have a life span and lifecycles. The system lifecycle refers to the different phases a system goes through in the course of its life. The typical lifecycle stages of a system are depicted in Figure 3.2.

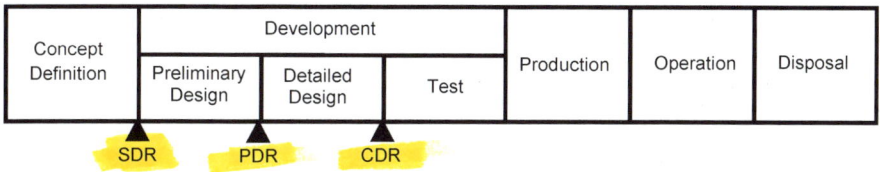

Figure 3.2 – System Lifecycle Phases

All aspects of a system can be characterized by what takes placed in each phase of this lifecycle model. HA is directed towards the operational phase, but it is in the development phase where hazards are created and controlled. The development phase is subdivided into Preliminary Design, Final Design and Test for more refinement. Under this model, each phase must be complete and successful before the next phase is entered, thus the system is developed in sequential stages. Three major design reviews are conducted for exit from one phase and entry into the next. These are the

System Design Review (SDR), Preliminary Design Review (PDR) and Critical Design Review (CDR). Different levels of HA are performed at each of these developments stages.

3.7 System Design

System design and development is an involved process. Depending on the type of system, the development process can take months or years. Design requirements are developed to show how the system is intended to work, and not work in the case of negative requirements. Design requirements are translated into design drawings, which show how the system actually works. Design functional diagrams show how the system is supposed to work, along with interrelationships. Software design requirements layout the way the system software is intended to work as it interfaces with the system hardware. Software code shows how the system software actually works. Procedures show how system users interface with the system and operate the system.

System safety engineering is a component of systems engineering and part of the system design and development process. System development goes through the phases of: *specify, design, build, and test* as shown in Figure 3.3. The information resulting from each of these processes are used by the HA analyst.

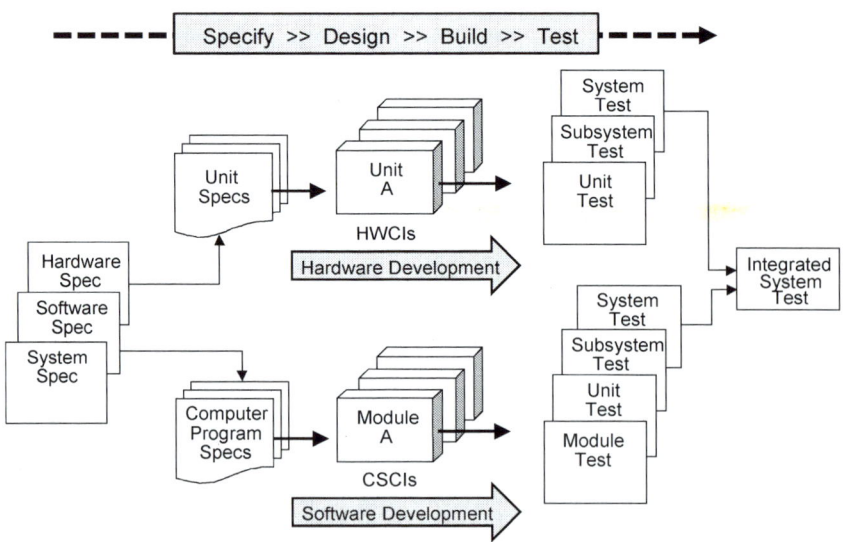

Figure 3.3 – System Development

3.8 System Hierarchy

Systems vary in size, shape, function, criticality and complexity. A system can be small, such as a toaster that consists of fewer than 50 parts. A system can be very large, comprising hundreds of subsystems, thousands of assemblies and millions of components, such as a commercial aircraft or a ship. Large complex systems can easily become overwhelming for human comprehension. In order to more easily visualize and understand a system, the system is typically broken down or subdivided in a hierarchical manner into manageable pieces that can be more easily grasped by the human mind. Referred to as a *system hierarchy table*[4], this view of the system defines the system structure in an orderly, linked and comprehensible manner.

The system hierarchy table establishes the organizational structure defining dominant and subordinate relationships between subsystems, down to the lowest component/piece part level. The system hierarchy table comprises a list of all the systems, subsystems, units, assemblies and components in the major system, with each item in the list indented to reflect its hierarchy and ownership level. The indenture level also identifies or describes the relative complexity of assembly or function. The levels progress from the more complex (system) to the simpler (part) divisions. System design data and drawings will usually describe the systems internal and interface functions beginning at system level and progressing to the lowest indenture level of the system.

The following is a typical breakdown of successive indenture level groupings in the system hierarchy:

- System – the system of interest
- Subsystem – the major subsystems comprising the system
- Unit – the major units comprising each subsystem
- Assembly – the assemblies comprising a unit
- Component – the components comprising an assembly
- Part – the lowest level of separately identifiable items, e.g., a bolt

Figure 3.4 demonstrates the hierarchy of elements in a system along with an example hierarchy table for an aircraft system.

---

[4] The hierarchy table also goes by other names, such as indentured equipment (IEL) list and master equipment list (MEL).

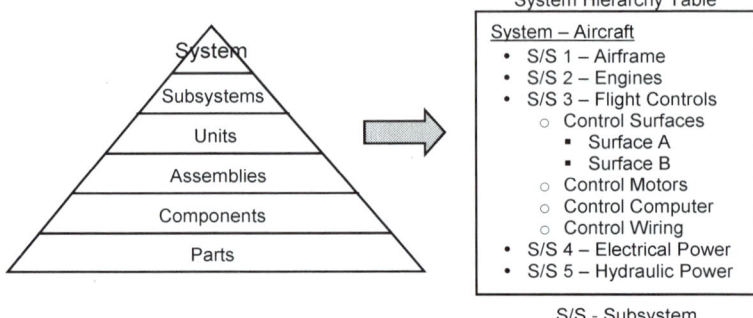

Figure 3.4 – System Hierarchy Representation #1

When performing a HA, all of the system components and functions must be considered to ensure a complete *systems* analysis. It is recommended that a system hierarchy table be established early in the program and used by the system safety practitioner in the performance of a HA. The following are some benefits gained from utilizing a system hierarchy table in safety analyses:

- Provides a complete list of equipment and functions
- Helps ensure that all of the system hardware and functions have been adequately covered by the HAs
- Helps set the level of detail for a particular safety analysis
- Helps establish at what level in the hierarchy hazards should be identified
- Helps in determining at what system level risk should be assessed

Another perspective of the system hierarchy is shown in Figure 3.5. This diagram shows that a typical HA must burrow down from the system level through the various intermediate levels until the contributing hazard component is reached and identified.

Figure 3.5 is a second viewpoint of system hierarchy. This view includes an example breakdown of both hardware and software. It does not, however, show the human interface with the system and the human tasks that must be performed.

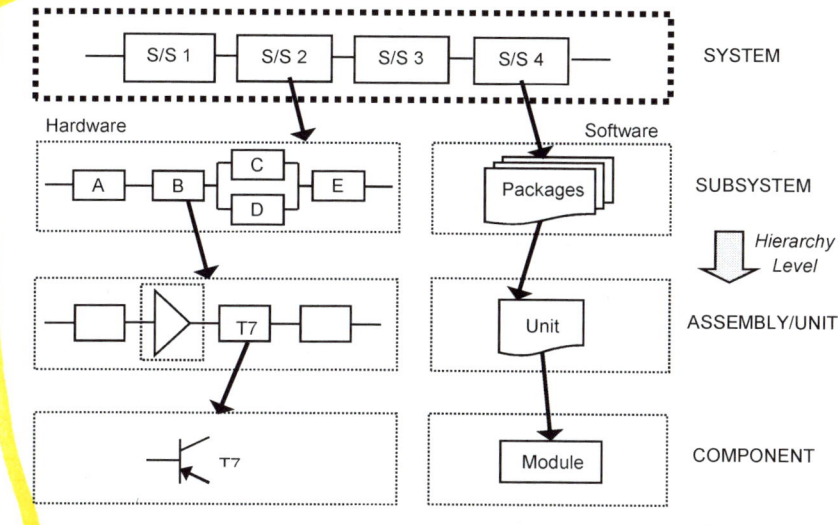

Figure 3.5 – System Hierarchy Representation #2

3.9 Systems Engineering Tools

In the process of developing large and highly complex systems, many tools are used by systems engineering and other related disciplines. Some systems engineering tools that greatly aid the system safety analyst in performing a HA include the following:

- Simplified System Diagram
- Functional Flow Block Diagrams (FFBD)
- System Hierarchy Table
- Reliability Block Diagram (RBD)
- Timeline Analysis
- Failure Mode and Effects Analysis (FMEA)
- System Function vs. Architecture Matrix
- Probability of Loss Table

Simplified system diagrams assist the safety analyst in understanding how a system operates, by viewing the subsystems and components within a system, along with their interface points. Figure 3-6 contains an example simplified system diagram of an unmanned aircraft system. From this diagram the safety analyst can perform a hazard analysis to identify system hazards, interface hazards and functional hazards. These diagrams are not meant to be detailed, but to show basic system content, interfaces and factors to consider and subsystem relationships. For sake of simplicity this diagram as shown here is not complete.

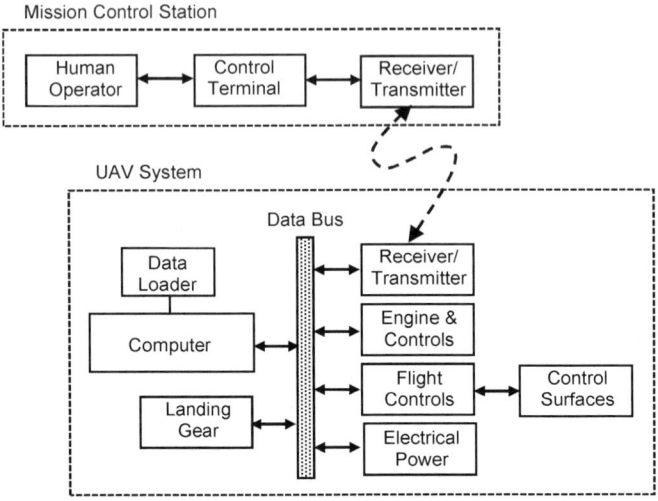

Figure 3.6 – Simplified System Diagram

A Functional Flow Block Diagrams (FFBD) layout is shown in Figure 3.7. The FFBD is a decomposition of system functions performed by the system. FFBDs provide the analyst with an understanding of how the system operates and which functions are safety-critical (SC).

The FFBD provides a means whereby individuals not familiar with the detailed design can quickly and easily grasp how the system operates. For example, a FFBD could be used to translate a complicated electric circuit diagram into a simplified series of functions.

FFBDs can also show hierarchy, in addition to system function. The top-level functions can be decomposed into lower level functions. And, each of the lower level functions can be decomposed into lower levels, and so on, until the bottom level is reached. The FFBD is also known as the Single Line Diagram (SLD), the Functional Flow Diagram (FFD or the Functional Dependency Diagram (FDD).

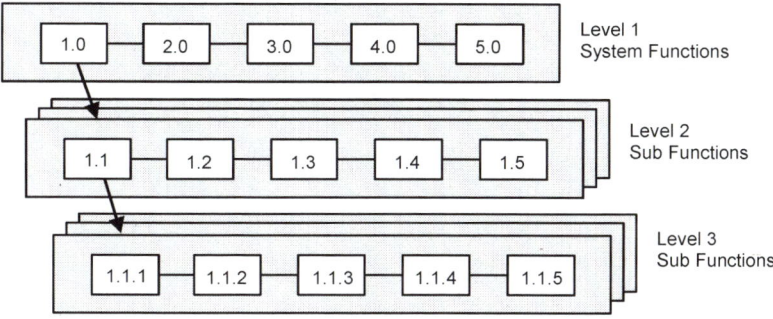

Figure 3.7 – Functional Flow Block Diagram (FFBD)

The system hierarchy table was presented in section 3.8. This table in invaluable in understanding the system design, and in ensuring that the HA covers all of the hardware and software in the system design.

The Reliability Block Diagram (RBD) is essentially a block diagram of the system components linked together in a reliability model. It shows what hardware and equipment are required for successful (reliable) operation of the system. It also shows the required combinatorial logic of how the components are linked together in the system design. RBDs are developed to fully understand system operation and to predict system reliability. An example RBD is shown in Figure 3.8. RBDs are typically developed by the program reliability organization. RBS provide design information and failure relationships important in HA.

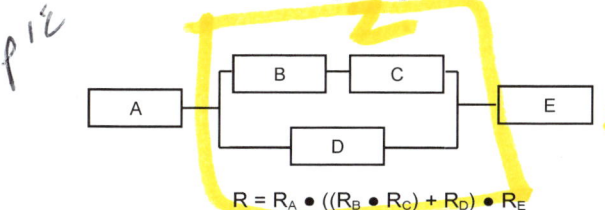

$$R = R_A \bullet ((R_B \bullet R_C) + R_D) \bullet R_E$$

Figure 3.8 – Reliability Block Diagram (RBD)

A system timeline analysis typically consists of a table of major system functions and components with their associated length of operation during system operation. It also shows when during the mission the components are operated. Timeline analyses define important and critical time sequences in the mission. A timeline analysis can also be in the form of an event sequence diagram that also includes event relationships and dependencies.

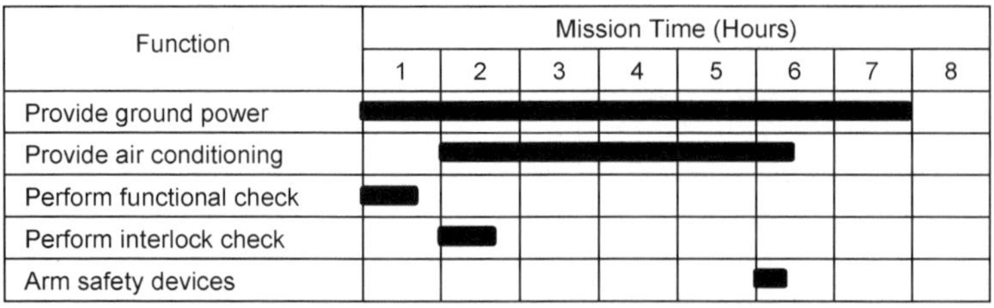

Function	Mission Time (Hours)							
	1	2	3	4	5	6	7	8
Provide ground power	■■■■■■■■■■■■■■							
Provide air conditioning		■■■■■■■■■■						
Perform functional check	■■							
Perform interlock check		■■						
Arm safety devices						■■		

Figure 3.9 – System Timeline Analysis

Failure Modes and Effects Analysis (FMEA) is an analysis tool for evaluating the effect(s) of potential failure modes of subsystems, assemblies, components or functions[5]. It is primarily a reliability tool to identify credible failure modes that would adversely affect overall system reliability. FMEA has the capability to include failure rates for each failure mode, in order to achieve a quantitative reliability analysis. FMEA is used to evaluate the reliability and safety effect of failure modes. When component failure rates are attached to the identified potential failure modes, a probability of subsystem or component failure can be derived. FMEA does not suffice for HA, however, failure modes identified in an FMEA are a good resource for hazard identification that can be effectively used during a HA.

Another useful HA tool is a function-architecture matrix shown in Figure 3.10. This matrix helps in understanding what system functions are performed in relation to the physical system hardware.

Functional Architecture	Physical Architecture				
Function	Airframe	Engine	Comms	Navigation	Fire Control
Pre-Flight	X	X	X	X	X
Load	X				
Taxi	X	X	X		
Takeoff	X	X			
Cruise	X	X	X	X	
Recon	X	X	X	X	
Communicate			X		

Figure 3.10 – System Function vs. Architecture Matrix

---

[5] See chapter 10 of reference 1 for more details on FMEA.

Figure 3.11 shows an example Probability of Loss table which is useful in performing a HA. This table shows which systems and subsystems contribute to Probability of Loss of Mission (PLOM), Probability of Loss of Aircraft (PLOA) and Probability of Loss of Control (PLOC) for an aircraft.

PLOM	PLOA	PLOC	Subsystem	Examples
↓	↓	↓	Processors	Computers, Data Buses, etc.
			Software	Aircraft control software
			Sensors	Air Data, Transducers, etc.
			Communication	VHF, UHF, etc.
			Navigation	GPS, INS, radar, etc.
			Flight Surface Actuation	Actuators, Servos, Valves, etc.
			Electrical Power	Generators, Batteries, Busses, Wires, etc.
			Hydraulic Power	Pumps, Valves, Filter, Pipes, etc.
			Fuel	Tanks, Valves, Lines, etc.
			Propulsion	Engine, FADEC, Oil Pumps, etc.
			Displays	Aircraft / ground operator displays
		↓	Data Link	Digital data link to unmanned aircraft
			Human	Pilot, Ground Operators for UAVs
			Environment	Rain, Ice, Temp, Vibration, Fire, EMI, etc.
			Operations	Exceeding envelope and margins
			Landing Systems	Landing Gear, Brakes, etc.
			Erroneous Commands	Inadvertent flight control commands
	↓		Flight Termination	Flight terminations systems
			Weapon Systems	Delivery, Stores Mgt., etc.
			Radar	Non-Navigational
			Reduced Redundancy	Loss of one redundant element
			Mission Unique	Tail Hook, Refuel, Lasers, etc.
↓			Lighting	Landing, Navigation, Anti-Collision, etc.

Figure 3.7 – Probability of Loss Table for an Aircraft

PLOA is a metric used primarily in reliability engineering to assess overall system level reliability of an air vehicle. It represents an estimate of the probability of loss or significant damage to the air vehicle under analysis over a given period of time. While it is usually expressed per hour, it can be calculated for any period of exposure. It should be noted that PLOA is also a metric used by system safety to represent the system level of safety for loss of an aircraft. Under the reliability aspect, PLOA typically means "Loss of Function", i.e., a failed or lost function affecting air vehicle system reliability. All subsystem functions that keep the air vehicle flying must perform properly under the reliability definition. System reliability is allocated to subsystems using the PLOA metric. Each major subsystem is assigned a probability of loss of function that it must meet in order to meet the desired system reliability when all subsystems are rolled-up to a top

PLOA number. Under the system safety aspect, PLOA includes many other possible causal factors in addition to Loss of Function, such as fire, collision, human error, and controlled flight into terrain.

PLOC is the estimated probability of degraded aircraft control due to system failures; PLOC includes only those systems and components necessary to maintain aerodynamic stability including speed, pitch, yaw and roll. Therefore, components such as landing gear would not be included. PLOC is a useful safety of flight (SOF) metric and can be used to assess airworthiness before flight test. PLOM is the estimated probability of mission failure due to failure of any system or subsystem. PLOM is the broader term, and is inclusive of PLOA and PLOC.

3.10 Summary

Systems are imperfect mechanisms developed to achieve an intended capability; unfortunately they often have unintended side effects when misused, abused or failures occur. When an item is evaluated for safety and reliability it must be evaluated by itself and also as it functions within the system. When an item is in a system it is exposed to a new set of dynamics and environments that it does not see separately from the system. These new dynamics present a new set of emergent properties involving hazards and risk. Figure 3.11 shows a model of a subsystem and a subsystem as it resides within the system. The subsystem is exposed to a special internal system environment comprised of many different internal interfaces and interactions. At the same time, the system is exposed to a special external environment comprised of many different internal interfaces and interactions.

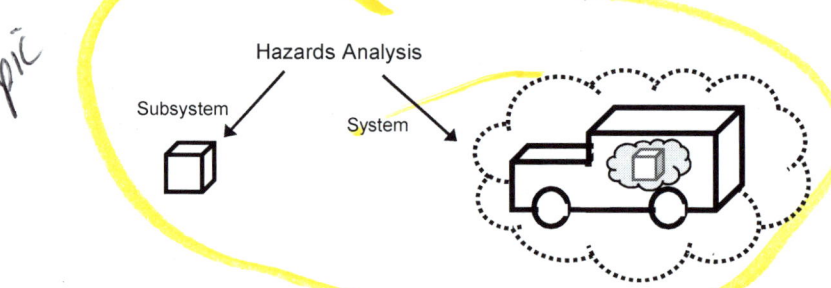

Figure 11.1 – HA Must Consider System and Subsystems

It is imperative that HA be performed on a subsystem and system basis, and that the many different environments and interface are considered. It is also important that the many system engineering tools are utilized in the conduct of a HA.

CHAPTER 4

HAZARD THEORY

4.1 Introduction

In order to identify and control hazards, it's necessary to fundamentally understand hazard theory. To successfully recognize a hazard it's essential to know what theoretically comprises a hazard; to eliminate a hazard it's necessary to know what specific hazard causal factors to confront. Hazards involve scenario, context and risk, which are based on the unique design aspects of a system. HA is a complex process that requires a firm grasp of hazard theory in order to avoid analyses resulting in poorly defined hazards, incorrect hazards and inconsistent hazard coverage.

Hazards, mishaps and risk are directly linked. Effective HA and hazard mitigation requires a thorough understanding of this conceptual relationship. This chapter focuses on hazard theory and establishes a viable hazard model (and definition) known as the HS-IM-TTO model, or just S-M-O in a shorter form. This hazard model helps to understand hazards, and provides a mechanism to identify them.

4.2 The Hazard-Mishap Relationship

The term *hazard* is a common, well used term; most people feel they understand it and that its definition is intuitively obvious. In its colloquial usage the definition is fairly simple, straightforward and broad. However, in the field of system safety the technical definition of a hazard is a little more complex, narrower in scope and possibly less well understood. In the past, the terms *hazard*, *risk*, and *danger* have been used interchangeably and incorrectly in the public domain, resulting in an incorrect understanding of safety.

Hazards and mishaps are directly linked. In order to understand the concept of a hazard, both terms must be understood and appropriately applied together. The following are typical definitions used within industry[6]:

> Hazard: Any real or potential condition that can cause injury, illness, or death to personnel; damage to or loss of a system, equipment or property; or damage to the environment.
>
> Mishap: An unplanned event or series of events resulting in death, injury, occupational illness, damage to or loss of equipment or property, or damage to the environment.

These are decent initial definitions, but they leave too much room for speculation and are difficult to use in exacting engineering applications. For example, using just this definition for a hazard, it is commonly assumed that gasoline is a hazard. However, if gasoline is a hazard, then what is the risk involved? In attempting to mitigate hazard risk it becomes obvious that more information is required than provided in the above definition, in order to effectively describe and understand a hazard. As will soon be learned, gasoline is a *hazard source*, but not a hazard.

For technical purposes, a hazard can be thought of as: <u>*an existing set of specific causal factors that form the potential for a mishap event*</u>. In other words, a hazard is an existing system state that is dormant, but has the potential to result in a mishap when the inactive hazard components are actualized. A hazard is a potential mishap, while a mishap is an event that has occurred as a result of a hazard that has become armed and active. This more technical definition is necessary because in order to mitigate the risk presented by a hazard all of the components and parameters the hazard comprises must be identified and understood.

Understanding that a hazard is a *potential* mishap is essential to understanding safety, as there is a direct link between the two. The metric of risk can be derived from the hazard components. Risk can be changed (mitigated), but only when the hazard components are known, understood and modified. Therefore, it is necessary to identify and understand the complete composition of a hazard, within the system context, in order to mitigate the hazard and the mishap risk it presents.

Another way to visualize the direct link between a hazard and a mishap is to compare the definitions of each. A hazard is defined as any real or potential <u>condition</u> that <u>can cause</u> injury, illness, or death to personnel; damage to or loss of a system, equipment or property; or damage to the environment. A mishap is defined as an unplanned <u>event</u> or

---

[6] MIL-STD-882D, *Standard Practice For System Safety*, Feb 2000.

series of events <u>resulting in</u> death, injury, occupational illness, damage to or loss of equipment or property, or damage to the environment. These definitions show that the outcome effect of each is the same – death, injury, etc. A mishap is an event resulting in loss and a hazard is a condition that can cause a loss event, thus, a hazard is a potential condition that can cause an actual mishap event.

A hazard is somewhat of a physical entity that characterizes a potential mishap; it is a condition that is prerequisite to a mishap; it is essentially a blueprint for a mishap. In order for a hazard to exist, three required components are necessary, which form a Hazard Triangle. A mishap is the result of an armed and actuated (triggered) hazard. Hazards are the result of hazardous system components, poor design and/or inadequate design foresight.

A hazard and a mishap are *before* and *after* states linked by a transition mode. The transition mode arms and activates the hazard. This concept leads to the principle that a hazard is the precursor to a mishap; a hazard defines a potential event (i.e., mishap), while a mishap is the event occurrence. This relationship between a hazard and a mishap are depicted in Figure 4.1.

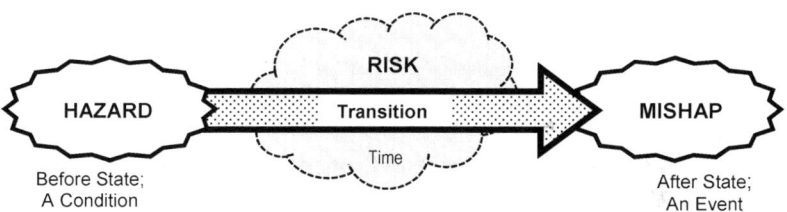

Figure 4.1 – Hazard / Mishap Relationship

A hazard and a mishap are two separate states of the same phenomenon, linked by a state transition. A hazard is a "potential event" at one end of the spectrum (before state), that transforms into an actual mishap event at the other end of the spectrum (after state). The state transition results from the occurrence of the hazard Initiating Mechanisms. A hazard is a latent condition that exists in a system, that when activated results in a mishap. Mishaps are the immediate result of actualized hazards. The state transition from a hazard to a mishap is based on two factors: 1) the unique set of hazard components involved and 2) the mishap risk presented by the hazard components. The hazard components are the items comprised by a hazard, and the mishap risk is the probability of the mishap occurring and the severity of the resulting mishap loss.

Hazard/Mishap risk is a fairly straightforward concept, where risk is defined as:

Risk = Likelihood x Severity

The mishap likelihood factor is the likelihood of the hazard components occurring and transforming into the mishap. Likelihood can be expressed in terms of probability, failure rate or qualitative ranges. The mishap severity factor is the overall consequence of the mishap, usually in terms of loss resulting from the mishap (i.e., the undesired outcome). Severity can be defined and assessed in either qualitative terms or quantitative terms. Time is factored into the risk concept through the probability calculation of a fault event, for example, $P_{FAILURE} = 1.0 - e^{-\lambda T}$, where T = exposure time and λ= failure rate.

Risk is an important component of the hazard-mishap relationship. Risk is a measurement that rates the overall safety significance (or danger) of a hazard. Risk is a measure that tells safety analysts how likely, or how often, the hazard will transform into a mishap. The risk measure also tells the safety analyst how severe the final mishap outcome is likely to become.

Risk management is an important tool for making critical decisions, and, in some cases, meeting regulatory requirements. In system safety, risk management is a process for the identification of hazards and their causes, determining the consequences of the hazards, calculating the probability of their occurrence and determining whether the risk is acceptable or if corrective actions are needed to make the risk acceptable. Hazard risk management is a key element of the system safety process.

However, risk management must not negate the system safety objective of eliminating hazards. No hazard should be accepted when that hazard can be reasonably eliminated or reduced in risk via design safety measures. Risk should be accepted only when the benefits outweigh the potential mishap damages, losses and costs. Risk information, risk knowledge and risk management are not an excuse for not eliminating hazards when feasible.

4.3 Hazard Characterization

A hazard is typically defined as a potential condition that can cause a mishap resulting in undesired consequences. From a more technical perspective, a hazard can be described as the existence of a specific set of system conditions that together create the possibility of a potential mishap; a hazard is an existing dormant system state, which has the potential to result in a mishap when the dormant hazard components are actualized. A hazard is a potential, while a mishap is an actual. A hazard can be thought of as a scenario comprising specific dormant causal factors, that when activated, result in a mishap.

4.3.1 Why Hazards Exist

Hazards occur in a system for one simple reason – the system must utilize, or interface with, a hazard source (hazardous asset) in order to achieve its intended goals. A mishap happens for two reasons – a hazard exists and the hazard was not eliminated or properly mitigated.

Hazards are created because of the need for hazardous sources in the system, or they must interface with hazard sources, coupled with the fact that eventually everything fails, and these failures can unleash the undesired effects of the hazard source. Hazards also exist due to the need for safety-critical system functions, coupled with the potential for failures and human error within these safety-critical functions. Hazard creation can be summarized by the following factors, which can occur singularly or in combinations:

- The use of hazardous system elements (e.g., fuel, explosives, electricity, velocity, stored energy)
- The system interfaces with hazard sources (e.g., fuel, electricity)
- Operation in hazardous environments (e.g., flood zones, ice, heat)
- The need for hazardous functions (e.g., aircraft fueling, welding)
- The use of safety-critical functions (e.g., flight control, arming)
- The inclusion of (unknown) design flaws, errors and sneak paths
- The potential for hardware wear, aging and failure
- Inadequacy in designing to tolerate critical failures

4.3.2 The Hazard Puzzle

A hazard is as a *potential condition that can cause a mishap, resulting in undesired consequences*. There is a direct link between a hazard and a mishap; a hazard being the precursor to a mishap. The definition of a mishap is fairly well understood, it is an event that has occurred with undesired outcomes, such as death, injury and/or damage. An industry problem is that the definition and concept of a hazard is not well understood or precisely defined, especially for technical purposes. For example, what specifically is a "potential condition" and how can it be understood and represented in order to consistently identify hazards? A condition can be many different things to different people. In addition, a hazard is often misunderstood because it is an actual existing condition with potential consequences. To fully understand, appreciate, recognize and measure hazards requires a more rigorous model based definition. It is no longer acceptable to leave the hazard definition open to individual interpretation.

A hazard is a real and existing condition with a potential undesired mishap outcome. It is recognized within the safety community that a hazard essentially has three states: dormant, armed and active. It is initially in an existing dormant condition, which is armed when it is activated by certain causal factors, and after becoming armed it transitions to the active state, which is the mishap event. The transition from armed to active may be immediate or it may take some length of time, depending upon the hazard components.

Hazard identification can no longer be an intuitive process, it must be a scientific process based on critical thinking. Several hazard models have been proposed; unfortunately many of them do not provide a consistent description of what actually comprises a hazard when it is in the condition state. It is necessary to understand the constituent components of a hazard in order to consistently recognize them and evaluate their risk and mitigation. A condition is defined as a state at a particular time; a state is defined as the way something is with respect to its main attributes (i.e., a situation). This infers that a condition is a concrete situation (or scenario) that can be clearly specified, with precise parameters.

Understanding that a hazard is a potential mishap, with specific causal factors, is essential to understanding safety. The metric of risk can only be derived from the hazard causal factors. Risk can be changed (mitigated), but only when the hazard components are known and understood. Therefore, it is necessary to identify and understand the composition of all hazards, within a total system environment, in order to understand and mitigate the risk before a mishap actually occurs.

The HS-IM-TTO hazard model makes the most logical sense in defining what comprises a hazard. This model breaks the hazard condition into three recognizable aspects or components: a hazard source (HS), an initiating mechanism (IM) and a target-threat outcome (TTO). All hazards require these three components in order to exist. This model supports the dormant hazard state, the causal factors that cause the hazard to become armed and the final outcome to expect when the hazard becomes active and transitions to a mishap. The HS is the element that causes the basic danger; the IM is the element that arms and activates the hazard; the TTO is the expected outcome severity when the mishap occurs. The hazard condition is a scenario that could result in a mishap, which is characterized by the HS-IM-TTO constituent components. A hazard description must contain the complete scenario context with all three of the required hazard components identified.

There is a lot of confusion about hazards, for example, one hazard model states that a hazard can be caused by other contributory hazards. This suggests a multi-hazard link where there is an initiating hazard and then contributory hazards. This is not a useful (or correct) concept. A hazard can, however, be caused by one or more causal factors (trigger events). Also, different hazards can be very similar with slightly different outcomes due to a slight change in one of the causal factors. This situation could be called a family of hazards. Trigger events are the hazard initiating mechanisms, such as failures, human error, etc. In the HS-IM-TTO hazard model, the IM component accounts for the trigger factors, whether or not they are referred to as initiating and contributory events (but they are not referred to as additional hazards). A hazard contains only the minimal causal factors that will cause it to occur. If additional extra causal factors can cause the hazard, then they form a separate, but similar, hazard. Bottom line – one hazard ... one minimal set of causal factors... one mishap.

4.3.3 The HS-IM-TTO Hazard Model

The HS-IM-TTO hazard model is the most logical for defining what comprises a hazard. In order for a hazard to exist, three hazard components must be present to form the hazard: 1) the Hazard Source (HS) which provides the basic source of danger, 2) the potential Initiating Mechanisms (IM) that will transition the hazard from an inactive state to a mishap event, and 3) the Target-Threat Outcome (TTO), or consequences, that will result from the mishap event.[7] The existence of a hazard requires the existence of these three components as a prerequisite. A hazard is an existing potential condition (inactive) that will result in a mishap when actualized. The hazard condition is a potential state formed by the Hazardous Source (e.g., energy) and the potential Initiating Mechanisms (e.g., failures) that will transform the hazard into a mishap and the hazard Outcome. The hazard Outcome is predicted in the expected mishap Target (e.g., personnel) and the expected mishap Threat (e.g., death or injury). These components are necessary to assess the risk and to know where and

---

[7] Adapted from work by Pat Clemens as described in System Safety Scrapbook, sheet 98-1, *Describing Hazards*, 1998, Sverdrup Corp. and work by J. Letos in Boeing document D180-28993-302, *Introduction to System Safety Analysis Process*, 1995, page 22.

how to mitigate the hazard. Table 4.1 describes these components in more detail. The HS-IM-TTO model is also referred to as the S-M-O model, for Source, Mechanism and Outcome.

Table 4.1 – Hazard Components

Hazard Component	Function or Purpose	Examples
Hazard Source (HS)	This is the element that provides the basic source of danger. Without it there would likely be no hazard.	• Energy sources • Safety-critical functions • Adverse environments
Initiating Mechanism(s) (IM)	These are the initiators or mechanisms that cause the mishap event to occur.	• Hardware failures • Human errors • Bent connector pins
Target-Threat Outcome (TTO)	This describes the potential outcome when the hazard becomes an actual mishap. There has to be a potential target and a threat to that target in the form of consequence and severity.	• Death or injury threat to humans • Damage threat to system or product • Damage threat to the environment

A more technical hazard definition is: A hazard is a set of dormant conditions, which consist of a Hazard Source, an Initiating Mechanism and a Target-Threat Outcome, which leads to a mishap when the Initiating Mechanism is actualized. A hazard is a physical entity that characterizes a potential mishap. A hazard is a condition that is prerequisite to a mishap, i.e., it is a blueprint for a mishap. When a hazard exists it forms a *Hazard Triangle*.

Basically, a hazard exists as a result of a system component being present in the system that presents safety vulnerability (i.e., the HS), combined with a system design that poorly tolerates various failure mechanisms that can affect the HS component. The actual amount of risk presented by the hazard is a function of how the design responds to the HS component when it is subjected to factors such as: hardware failures, human errors, sneak electrical paths, software errors, incorrect interfaces, etc. In order to mitigate a hazard, the hazard must be recognized and understood, and then the influence factors appropriately modified. Table 4.2 provides some example items and conditions for each of the three hazard components.

Table 4.2 – Hazard Component Examples

Hazard Source (HS)	Initiating Mechanism (IM)	Target/Threat Outcome (TTO)
• Ordnance • High pressure tank • Fuel • High voltage	• Inadvertent signal; RF energy • Tank rupture • Fuel leak and ignition source • Touching an exposed contact	• Personnel; Explosion; Death/Injury • Personnel; Explosion; Death/Injury • Personnel; Fire; Death/Injury • Personnel; Electrocution; Death/Injury

To demonstrate the hazard HS-IM-TTO model, consider a detailed breakdown of the following example hazard: "Worker is electrocuted by touching exposed contacts in electrical panel containing high voltage." Figure 4.2 shows how this hazard is divided into the three necessary hazard components to validate the hazard. Note in this example that all three hazard components are present and can be clearly identified. In this particular example there are actually two IMs involved. The TTO defines the mishap outcome, while the combined HS and TTO define the mishap severity. The HS and IM are the hazard causal factors that are used to determine the mishap probability. If the high voltage component can be removed from the system, the hazard is eliminated. If the voltage can be reduced to a lower less harmful level, then the mishap severity is reduced and the hazard is mitigated to a lesser level of risk.

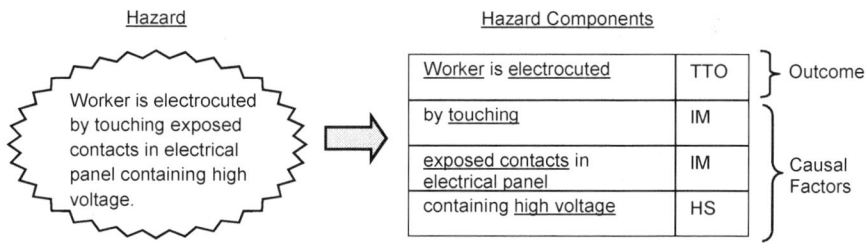

Figure 4.2 – Example of Hazard Components

A hazard is an existing potential condition (inactive) that will result in a mishap when actualized. The hazard condition is a potential state formed by the Hazardous Source (e.g., energy) and the potential Initiating Mechanisms (e.g., failures) that will transform the hazard into a mishap and the hazard Outcome. The hazard Outcome is predicted in the expected mishap Target (e.g., personnel) and the expected mishap Threat (e.g., death or injury).

Figure 4.3 depicts the S-IM-TTO hazard model with more detailed examples of possible system factors that could produce each of the S-IM-TTO categories. This example list of factors is not complete or exhaustive, but it becomes apparent that the list of HS factors can be very extensive. Systems use many HSs in their design; these HSs must be dealt with via the system safety HA process in order to produce safe systems and products.

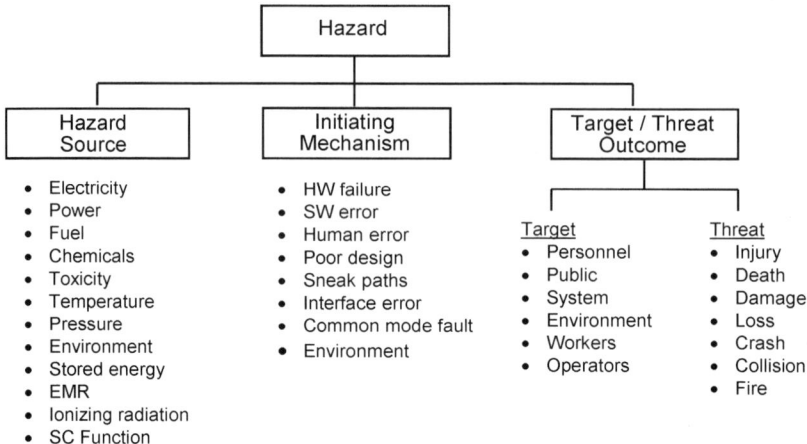

Figure 4.3 – Expanded Hazard Model

4.3.4 Hazard Triangle

The three required components of a hazard form what is known as the Hazard Triangle, which is illustrated in Figure 4.4. The Hazard Triangle conveys the idea that a hazard consists of three necessary and coupled components, each of which forms the side of a triangle. All three sides of the triangle are essential and required in order for a hazard to exist (i.e., HS, IM and TTO). Remove any one of the triangle sides and the hazard is eliminated because it is no longer able to produce a mishap (i.e., the triangle is incomplete). For example, remove the human operator from an aircraft, and the hazard "aircraft crashes resulting in pilot death" is eliminated because the TTO is removed. Reduce the probability of the IM side and the mishap probability is reduced. Reduce an element in the HS or the TTO side of the triangle and the mishap severity is reduced.

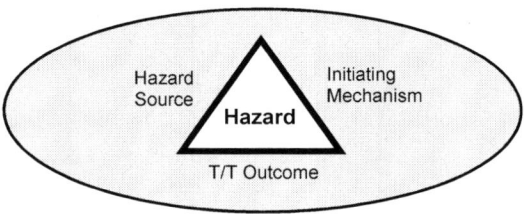

Figure 4.4 – Hazard Triangle

In the Hazard Triangle, the Hazard Source is the system hazardous asset. It sets the stage for the hazard. The Initiating Mechanism involves the basic factors that take the hazard from a potential to an actuated event. The Target-Threat Outcome is the basic danger presented by the hazard. It defines who or what is targeted for an undesired event and what the specific threat involves.

A hazard can be eliminated by eliminating any one of the three basic components of a hazard (HS, IM or TTO) because this breaks the hazard component coupling. Practically speaking, however, hazards are primarily eradicated by eliminating the HS component. Hazards are predominantly mitigated in risk by reducing the probability of the IM. Hazard risk can also be mitigated by reducing the effective danger of the HS, by protecting the Target or by reducing the amount of the Threat. It should be noted, however, that most hazards are mitigated by reducing the actuating probabilities of the IM. Changes in the probability and/or severity of a hazard modify the risk of the hazard. The Hazard Triangle concept is useful in determining if a conceptualized hazard meets the necessary criteria and in determining where to mitigate a hazard. It also demonstrates that when a hazard is mitigated it is not eliminated (a common error in thinking of safety beginners), because all three sides are still present. It's noteworthy that these mitigation methods correlate appropriately with the Safety Order of Precedence.

4.3.5 Good Hazard–Bad Hazard

Correctly describing the hazard is a very important aspect of hazard theory and hazard analysis. The hazard description must contain all three components of the hazard (Hazard Source, Initiating Mechanism, and Target/Threat Outcome). The hazard description should also be clear, concise, descriptive and to the point.

If the hazard description is not properly worded and clearly understood, it will be difficult for anyone other than the original analyst to completely understand the hazard. If the hazard is not clearly understood, the concomitant risk of the hazard may not be correctly determined. This can lead to other undesirable consequences, such as spending too much time mitigating a low-risk hazard, when it was incorrectly thought to be a high-risk hazard. Table 4.3 shows some good and poor examples of hazard descriptions.

Note that the good description examples contain all three elements of a hazard (HS, IMs and TTO). It should also be noted that the poor hazard description examples are primarily hazard sources only, or they are pseudo hazards. It should be emphasized that occasionally the HS and IM components may overlap or their distinction is blurred. In these cases it is important that all the components are identified, and divisive arguing should not be spent on whether or not they are labeled specifically as HS or IM factors.

Table 4.3 – Example Hazard Descriptions

Poor Examples	Good Examples
Broken glass	Worker accidentally breaks glass window and severely cuts himself from the broken glass.
High voltage	Worker is electrocuted by touching exposed contacts in electrical panel containing high voltage.
Gasoline	Automobile is hit from the rear by another auto, causing the fuel system to rupture; spilled fuel is ignited, resulting in fire that severely injures occupants.
Repair technician slips on oil.	Overhead valve V21 leaks oil on walkway below, spill is not cleaned, repair technician walking in area slips on oil and falls on concrete floor, causing serious injury.
Signal MG71 occurs.	Missile Launch signal MG71 is inadvertently generated during standby alert, causing inadvertent launch of missile and death/injury to personnel in the area of impacting missile.
Round premature	Artillery round fired from gun explodes or detonates prior to safe separation distance, resulting in death or injury to personnel within safe distance area.
Ship causes oil spill	Ship operator allows ship to run aground, causing catastrophic hull damage, causing massive oil leakage, resulting in major environmental damage.

4.3.6 Hazard Context

Hazards must be described in a proper and complete systems context. This context must explain the complete system scenario and causal factors, which include the HA, IM and TTO hazard components. Do not abbreviate the hazard description or use program-unique lingo, assuming the reader fully understands the text. Describe the hazard scenario in a meaningful context that fulfills the three sides of the Hazard Triangle. For example, "fuel" is not a hazard, it is a hazard source. "Fuel leak occurs due to *xxx* and an ignition source occurs due to *xxx*, leading to fire and system loss" is a hazard. In this example, the actual causes of the fuel leak and the ignition source must be included. If the system context of the hazard is not fully described, then the hazard causal factors and hazard risk cannot be established, because a risk calculation requires something concrete from which to derive likelihood and severity.

4.7 Hazard Laws

Some natural laws or truisms relative to hazard theory, which aid in the HA process, include the following:

1) Mishaps are not just random chance events; they are the result of actualized hazards existing in the system design.
2) The definition of a hazard is more complex than generally realized. Most definitions are inadequate; a more technical and specific definition is required for engineering purposes.
3) A hazard is a real entity that exists within the design of a system; it is formed by the unique features and components of a system.
4) Hazards are man-made, created during the design of a system. They are the result of flawed designs, hazardous assets and potential component failures.
5) A hazard exists in a dormant state until it is activated, whereby it then transitions into a mishap.
6) Hazards either exist or they do not, there is no in between or maybe.
7) A hazard is comprised of three necessary and required components: hazard source (HS), initiating mechanism (IM) and target/threat outcome (TTO).
8) In order to identify a hazard, the three components of a hazard must be identified: the HS, the IM and the TTO. It is only when these three components have been established and defined, that a hazard has been fully identified.
9) The HS-IM-TTO components form a Hazard Triangle. In order for a hazard to exist all three sides of the triangle must be present. Removing any one component from the system design removes one side of the triangle, thereby eliminating the hazard. Mitigating any one side of the triangle, via system design, reduces the risk presented by the hazard.
10) Hazard risk equals mishap risk. A hazard defines a potential future event, whereas a mishap defines an occurred event. It's the same risk, just from two different viewpoints; they each involve the same causal factors and outcome

11) Hazard characterization must include enough information to establish the risk involved and the factors that can be mitigated. This requires establishing the hazard scenario and context, including the HS, IM and TTO components.

12) In many cases the system design has a need for a hazardous asset, such as fuel, high voltage, moving parts, etc. Whenever a hazardous asset exists within a system, there will always be one or more unique hazards spawned by that asset.

4.6 Summary

After reviewing the hazard, mishap and risk concepts presented in this chapter, I propose the following definitions:

<u>Hazard</u>

A hazard is an existing set of conditions that form the potential for a mishap; a threat of harm [short form].

A hazard is a precondition to a mishap; it is an existing set of specific conditions forming the potential for a mishap, that when activated becomes the actual mishap event. The conditions comprising a hazard consist of three required elements: a Hazard Source (HS), an Initiating Mechanism (IM) and a Target-Threat Outcome (TTO). The IM can consist of multiple combined mechanisms, or sub events [long form].

<u>Mishap</u>

A mishap is an unplanned event resulting in harm; the result of an activated hazard [short form].

A mishap is an unplanned event resulting in death, injury, occupational illness, damage to or loss of equipment or property, or damage to the environment. Overall it is a single final event; however, it may be comprised of unique sub events that together activate the mishap, such as hardware failures, human error, environmental conditions, etc. [long form].

CHAPTER 5

HAZARD RISK THEORY

5.1 Introduction

As was shown in chapter 4, the relationship between hazards and mishaps involves the element of risk. Risk is a safety aspect that can be measured and controlled. Risk is not a tool for avoiding safety action; it is a tool for determining when action is necessary and when that action has adequately achieved its safety goal. This chapter focuses on the concept of risk, and how risk is used in the hazard mitigation (risk reduction) process.

5.2 Risk

The concept and reality of risk has been around for some time. There are many different types of risk, such as: safety risk, hazard risk, mishap risk, schedule risk, cost risk, investment risk, product risk and sports risk, just to name a few. Risk also involves many different viewpoints, such as: perceived risk, real risk, individual risk, group risk, societal risk, high risk takers, low risk takers, risk aversion, etc. On the surface, risk appears to be a very simple concept; however, risk can easily become very complex due to all the types, factors, possibilities and considerations involved. Risk and risk management are not just safety concepts; they are used in many different fields, such as finance, project management, health care, etc. Risk is not about the present, it is about the future. Risk deals with potential future events, and the uncertainty and outcomes of these events.

Risk is a vector value combining event likelihood with event outcome. It is a metric expressing the expected value of a future event based on the parameters creating the potential event. It expresses the likelihood of a potential gain/loss from a given decision. Risk involves three parameters: a) a potential future event, b) the likelihood of the event occurring and c) the expected consequences from the event when it occurs. Each of these aspects involves an element of uncertainty. Risk is defined as the product of the event likelihood and the event outcome, where the outcome can be either a positive or negative consequence, depending on the event. Risk outcome is the final expected result of the future event, given that it occurs. Risk outcome can be a threat or an opportunity; however, in system safety it is treated as a threat of loss, damage, death, injury or any combination of these outcomes.

Risk is an intangible quality; it does not have physical or material substance (a mishap does, but not risk). It is a future-value concept with some quantifiable metrics, likelihood and severity, that characterize the future event. In system safety, risk is a measure of the future event, where the event is an expected mishap. Risk likelihood can be characterized in terms of probability, frequency or qualitative criteria, while risk severity can be characterized in terms of death, injury, damage, dollar loss, etc. Recognizing that a hazard is the precursor (or blueprint) of a mishap, safety risk is the common denominator between the hazard and a mishap, and also the measure of the relative threat presented by a hazard.

Risk is a method for quantifying danger and the uncertainty involved. Some basic axioms regarding safety risk include the following:

- Risk is a metric of the likelihood and consequence of a potential future event
- In the case of safety, the future risk event is a potential mishap
- Risk is a measure of the predicted amount of danger presented by a hazard, in terms of a future mishap
- When a hazard exists, there is always risk associated with it
- Hazards and their components must be identified before risk can be assessed
- Hazard risk varies based on the hazard components involved
- Risk is in effect regardless of whether it's known or unknown
- Risk can be eliminated, reduced or increased through design measures

5.3 Hazard Risk or Mishap Risk?

Safety risk is sometimes stated two different ways by individuals with different backgrounds, which can lead to confusion. The two types of safety risk terms are:

- Hazard risk – the safety metric characterizing the amount of danger presented by a hazard, where the likelihood of a hazard occurring and transforming into a mishap is combined with the expected severity of the mishap outcome predicted by the hazard.

- Mishap risk – the safety metric characterizing the amount of danger presented by a potential mishap, where the likelihood of the mishap's occurrence is combined with the resulting severity of the mishap. Mishap risk likelihood defines the likelihood of the mishap occurring, while mishap risk severity defines the expected final consequences and loss outcome expected from the mishap event. The mishap likelihood and severity can only be computed from the information contained in the hazard description, when the mishap is being predicted in advance.

It should be noted that hazard risk and mishap risk are really the same entity, just viewed from two different perspectives. Figure 5.1 depicts the hazard-mishap risk relationship and their common link – risk.

Figure 5.1 – Hazard Risk vs. Mishap Risk

Hazard analysis looks into the future and predicts a potential mishap from a hazard perspective. A potential mishap can only be recognized and understood in terms of a hazard; therefore, mishap risk can only be reached by first determining the hazard and then evaluating its risk in terms of a mishap from the hazard causal factors. Since a hazard merely pre-defines a potential mishap, the risk has to be the same for both a hazard and a mishap.

5.4 Risk Characterization

Risk characterization is important, because it provides the mechanism by which a system safety program can judge the criticality of hazards, and also judge the effect of hazard mitigation. Two significant questions in the risk acceptance process are: how should hazard risk be characterized for criticality judgment and what acceptance criteria should be used. The risk acceptance method selected must address the concern of complexity versus utility. If the judgment criteria method is too complex it will not be used effectively.

There are two basic approaches for characterizing risk: qualitatively and quantitatively. Qualitative methods generally involve establishing a subjectively defined risk table comprised of several risk categories, and then assigning each hazard into the appropriate risk division. Qualitative methods are actually very efficient and effective for evaluating the risk of hazards. There are several different methods for qualitatively characterizing risk. It is important to select the appropriate method, because each of the various approaches presents both advantages and disadvantages. Refer to reference 1 for some example qualitative methods.

Figure 5.2 shows a qualitative risk acceptance approach using the Hazard Risk Index (HRI) method recommended in MIL-STD-882. This method provides a good characterization of risk, which can be estimated qualitatively or semi-quantitatively. It also provides a relatively simple methodology that is cost-effective and time-effective to perform.

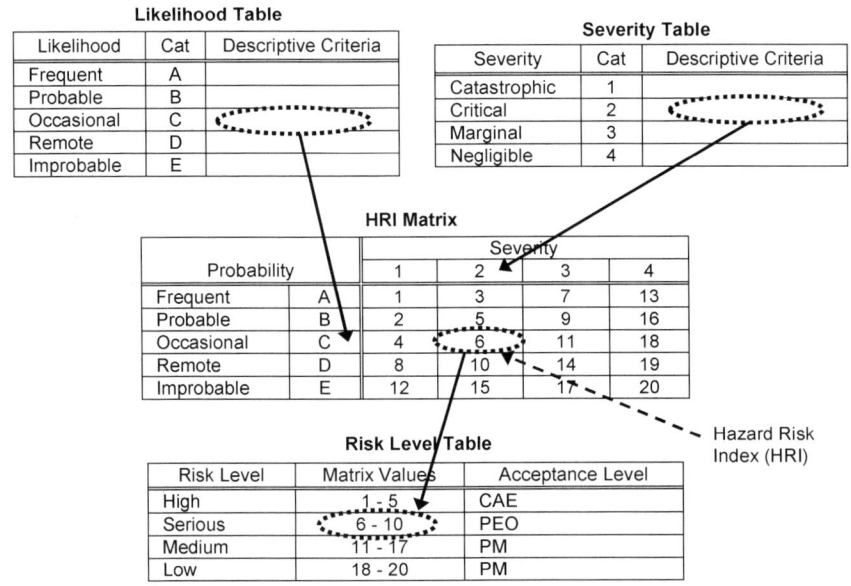

Figure 5.2 – Hazard Risk Index Matrix Concept

The HRI approach from MIL-STD-882 is the most commonly used approach. Later versions of MIL-STD-882 refer to HRI as Mishap Risk Index (MRI), which is effectively the same thing. See reference 2 for more detail.

The HRI matrix establishes the relative level of potential mishap risk presented by an individual hazard. By comparing the calculated qualitative severity and likelihood values for a hazard against the pre-defined criteria in the HRI matrix, a level of risk is determined by a derived index number. The HRI matrix concept essentially involves one matrix and three tables, as depicted in Figure 5.2.

The relative risk index (or HRI) is derived from the matrix cell resulting from the intersection of the likelihood and severity axes that represent a particular hazard. The HRI matrix maps hazard severity on one axis and hazard likelihood on the other axis. Once a hazard's severity and likelihood are determined, they are mapped to a particular HRI matrix cell, which yields the hazard index and the relative risk for that hazard. The hazard risk level establishes who can accept the risk by authority level.

Each matrix cell contains an HRI risk index number, indicating the relative (vice absolute) safety mishap risk presented by a particular hazard. Note that the cells for a 5 x 4 matrix are labeled with a risk index of 1 through 20, where 1 represents the highest risk and 20 the lowest risk. The smaller the HRI number, the higher the safety risk presented by the hazard. The HRI indices are divided into four groups which comprise the High, Serious, Medium and Low risk levels in the Risk Level Table.

Quantitative methods generally involve the use of probability and statistics for stating the likelihood of a hazard or mishap. Good quantitative risk measures are very time consuming and require considerable data and analysis. Fault Tree Analysis (FTA) is an effective tool for quantitatively predicting the probability of an undesired event or mishap. The HRI matrix could be modified such that the likelihood categories are divided into probability ranges, rather than qualitative ranges. Hazard probabilities can be computed by determining the probability of each causal factor in the hazard, and then multiplying these probabilities together.

Several different methods for evaluating risk are suggested in reference 11. A more advanced risk characterization method is the Iso-Risk contour, shown in Figure 5-3. The two variables that constitute risk define a risk plane, often referred to as an Iso-Risk contour[8]. In this approach the

---

[8] The Iso-Risk Contour, System Safety Scrapbook, page 84-5, Pat Clemens, Svedrup Corp.

likelihood and severity ranges can be expressed either qualitatively or quantitatively.

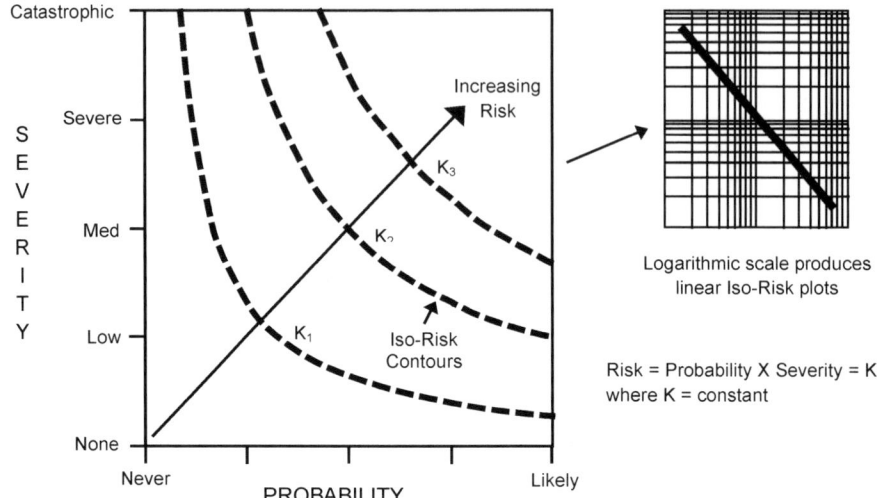

Figure 5.3 – The Iso-Risk Plane

The Iso-Risk contour shows the range of severity vs. probability combinations for which risk is constant. This characterizes risk for many hazards, but the analyst must be careful because there are exceptions. This model represents certain hazards that may have varying degrees of severity; the product of severity and probability for the various outcomes of the same hazard are relatively constant. The hazard may have a low severity with high probability or a high severity with low probability. This is a confusion factor that would seem to indicate a family of hazards rather than a single hazard. It is important to note that not all hazards behave this way. This figure shows the generalized Iso-Risk contour and its linear conversion. When the risk of a hazard is reduced (i.e., changed), the constant Iso-Risk plane changes to a new value.

5.4 Risk Reduction

The purpose of HA is to identify hazards and their associated risk, and then to eliminate the hazard (and its risk) or mitigate the hazard and thereby reduce its risk. Risk reduction involves applying a risk management process.

Hazard Analysis Primer

Measurement provides a mechanism for understanding an entity, and it also provides a means for evaluating changes to that entity. As Lord Kelvin is said to have stated "Anything that exists, exists in some quantity and can therefore be measured." Even though a hazard is a potential future event, its current value can be measured using the parameters of risk – likelihood and severity. Risk likelihood is the measure of the future event occurring, whereas risk severity is the measure of the amount of undesired consequence resulting from the future event when it occurs. Overall, risk is the metric characterizing the amount of danger presented by a hazard, thus risk is a key metric for measuring safety.

As shown if Figure 5.4, risk reduction involves several steps. First, the hazard and potential mishap must be identified. Next, the hazard-mishap risk must be calculated from the hazard causal factors. Finally, unacceptable risk must be reduced by applying design safety measures (i.e., defenses) to mitigate the hazard. The risk remaining after mitigation is referred to as *residual risk*.

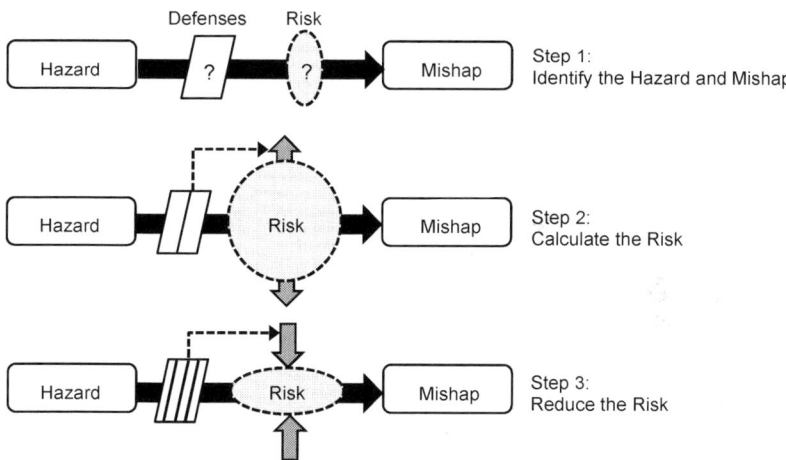

Figure 5.4 – Risk Reduction

In step 1, the hazard defenses (barriers) and risk are unknown. In step 2, the existing defenses are identified (if any); the defenses and hazard causal factors (HCFs) establish the risk. In step 3, defenses are added to the design, thereby reducing the risk.

Figure 5.5 shows the overall system risk breakdown following hazard mitigation and risk reduction. If a good HA is performed all hazards should be identified and the unidentified risk will then be minimal or nonexistent. If the HA is inadequate, then the unidentified risk (unidentified hazards) could be a concern.

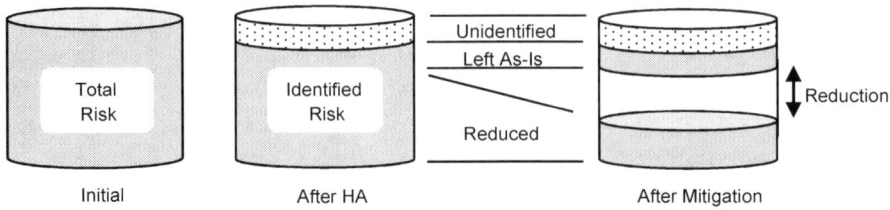

Figure 5.5 – Risk Breakdown Following Hazard Mitigation

There are two ways to mitigate the risk presented by a hazard: a) reduce the probability of the hazard occurring and b) reduce the severity of the hazard-mishap outcome, as depicted by Figure 5.6. In reality, it is usually very difficult, if not impossible, to change the potential severity outcome of most hazards. Therefore, the primary way to reduce risk is to attack and counter the hazards causal factors and reduce their likelihood of occurrence. Some risk may be acceptable and reduction is not necessary.

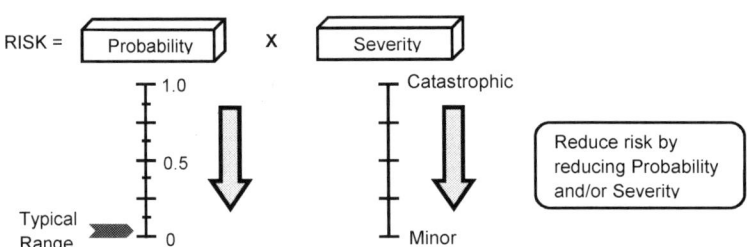

Figure 5.6 – Risk Reduction

5.5 Safety Order of Precedence (SOOP)

When developing DSF options to mitigate hazards, it is recommended that the Safety Order of Precedence (SOOP) be followed. This safety hierarchy protocol was originally established by MIL-STD-882. The SOOP provides a preferred order for implementing different DSFs, each of which affects the level of risk differently; each SOOP level provides a higher level of safety assurance in a step-like manner. Figure 5.7 graphically displays the SOOP. Note that design for minimum risk is the most desired SOOP method, after design for hazard elimination.

Figure 5.7 – Safety Order of Precedence

Hazards are typically mitigated following the Safety Order of Precedence (SOOP) for the most effective results. See reference 2 for a detailed description of the SOOP methodology.

The theory behind the SOOP is to first eliminate the hazard through design measures. For example, replacing an explosive device with a compressed gas cylinder (here the original hazard is replaced with another less severe hazard). If the hazard is eliminated, there is no mishap risk involved for that particular original hazard. However, many systems are inherently hazardous, such as weapon systems, and hazard elimination is not an option. In this situation, Branch B must be taken, whereby the risk is mitigated through various design options. Following Branch B, the first and most preferred option, is to reduce mishap risk through design measures (e.g., incorporate several safety interlocks into a weapon arm function). The next best option is to incorporate safety devices and/or safety features into the design that will reduce the mishap risk (e.g., thermal protection). If this is not possible, then utilize a detection and warning system to alert personnel to the particular hazard (fire alarms in the storage area). The last and least desirable option to reduce mishap risk is through the use of special procedures in the operation manuals, and/or special training for the personnel involved. Procedures may include the use of personal protective equipment. Procedures and training are not sufficient as the only risk reduction method for Category I and II hazards.

It should be noted that hazard mitigation does not necessarily have to be limited solely to one of the options in the safety order of precedence. Incorporating one or more of the order of precedence options may be utilized to mitigate a hazard. However, the best and recommended option is to always attempt to reduce mishap risk through design measures, and then utilize other measures as necessary.

5.6 Risk Management

Risk is a measure of the uncertainty associated with future potential events, in order that decisions regarding these events can be made today. Risk decisions can result in either negative or positive outcomes. In system safety, risk management involves the process involved in eliminating and mitigating hazards. Risk management is a tool used to make decisions in the present that will help to produce a desired safe outcome in the future.

Risk is a measure that rates the relative safety significance (or danger) of a hazard. Hazards and risk exist regardless of our perception, knowledge or awareness of their presence. Hazards and risk do not care if we know about them or try to do anything about them. John Adams summed it up nicely: "Risk management: it's not rocket science − it's much more complicated"[9].

The process of accounting for and addressing risks is called Risk Management. An example would be designing an automobile with air bags to reduce the risk of personnel injury should an accident occur.

Risk reduction involves applying a risk management process, which includes:

- Risk measurement
- Risk judgment
- Risk mitigation
- Risk acceptance
- Mitigation verification

Risk Management is one of the core elements of the system safety process; it is a key element in developing a system that presents acceptable mishap risk. An effective mishap risk management process must be formulated for each program and accepted by all of the stakeholders involved.

---

[9] John Adams, *Risk Management: It's Not Rocket Science − It's Much More Complicated*, Journal of System Safety, Mar-Apr, 2008, Pages 15-17.

There is risk associated with every hazard. A safety risk assessment must be performed on all identified hazards to determine the level of potential mishap risk they present. Rules for judgment of risk must be established to determine if the risk is acceptable or unacceptable. When the risk presented by a hazard is deemed unacceptable, it must then be eliminated or reduced.

Two significant questions in the risk acceptance process are, how should hazard risk be characterized for acceptance judgment and what acceptance criteria should be used. The risk acceptance method selected must address the concern of complexity versus utility. If the judgment criteria method is too complex it will not be used effectively. Typically the risk rating method and criteria are established prior to performing a HA and are documented in the System Safety Program Plan (SSPP).

When the risk presented by a hazard must be mitigated, it is typically done through the implementation of a design safety feature (DSF), such as a redundant component, fail-safe mechanism, interlock, etc. System Safety Requirements (SSRs) are the vehicle that establishes the DSFs for the system design. By implementing the DSFs contained in the SSRs, the mishap risk is reduced to an acceptable level. But, it does not end there; the SSRs must be verified and validated (V&V) to ensure they are implemented and effectively eliminate the hazard or reduce it to an acceptable level of risk. Mitigation success is verified through appropriate analysis, testing, demonstration or inspection. The use of a traceability matrix provides a trace from the hazard to the mitigating SSRs, and then to the test requirements, to the test results.

Risk reduction methods used in system safety may target reduction of the severity of the consequences of a hazard, the probability of the mishap's occurrence or both. Measures such as double hulls, fire-suppression and life-saving equipment are intended to reduce the consequences in the event of a mishap. Increased training to improve crew competency could reduce the likelihood of a mishap. Equipping the ship with the most advanced navigational tools is another risk management approach, however, it may prove counterproductive (increase risks) if proper training and procedures are not provided along with the enhanced technology. The key to successful risk management tools is that they be effective, specific to the hazards that create the risk and that they are relatively easy to implement. This will increase the likelihood of successfully mitigating the hazard and the risk.

Figure 5.8 depicts the overall risk management process for risk reduction. This process includes risk identification, risk evaluation, risk judgment, risk mitigation if necessary, risk mitigation verification and risk acceptance.

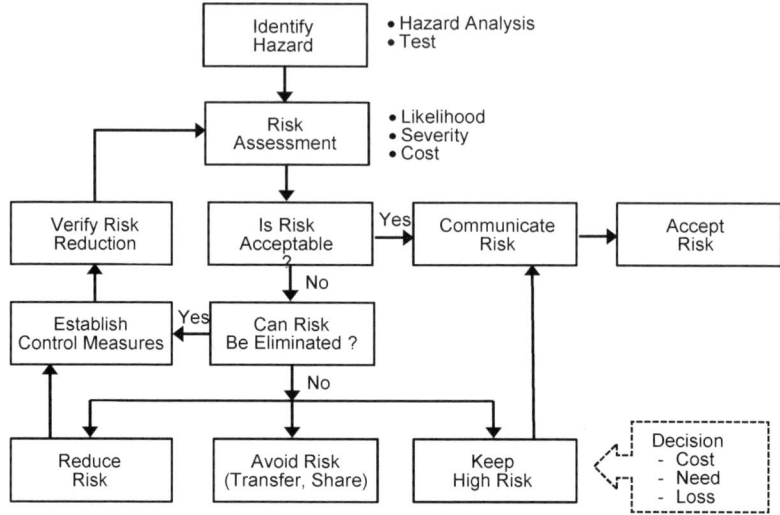

Figure 5.8 – Risk Management Process

A step often overlooked in the risk management process is mitigation verification. After control measures are established and implemented to mitigate the risk, it is necessary to verify that the proper steps were taken and the risk has truly been mitigated. Risk acceptance should never occur until the verification step has been successfully passed.

Another aspect of risk control is risk sensitivity. Sensitivity addresses the ability of the risk management approaches to affect the risk in the most effective places. For maximum impact, risk management control efforts should be directed toward those items driving risk, which are sensitive to intervention. For example, suppose the failure of a specific resistor, and a specific diode, cause a hazard. Sensitivity analysis would establish that redesign of the resistor would provide an 80%reduction in the risk, while the diode would only provide a 20% reduction.

5.7 Common Problems in Implementing Risk Controls

From a review of lessons learned on risk reduction controls, many risk controls never achieve their full potential. The primary reason for shortfalls is failure to effectively involve the personnel who are actually impacted by a risk control. The following is a list of problems commonly encountered when risk controls are implemented:

- Inappropriate for the problem
- Operators dislike it
- Managers dislike it
- Too costly (unsustainable)
- Overmatched by other priorities
- Misunderstood
- Progress measured too late
- Insufficient or inadequate documentation and training

5.8 Risk Probability

Hazard probability and severity depend upon different aspects of the system design. The risk probability calculation only has meaning when associated with an operating duration or exposure time. Increasing exposure time increases risk. An example of this fact in real life is auto insurance; the premium for one year insurance coverage is much more than for one month of coverage.

Reliability mathematics are used to calculate the probability of failure for the basic component failures in a HA. The basic reliability equations that are used in FTA are:

- $R = e^{-\lambda T}$ << this is the probability of success
- $R + Q = 1$
- $Q = 1 - R = 1 - e^{-\lambda T}$ << this is the probability of failure
- Approximation: $Q \approx \lambda T$ when $\lambda T < 0.001$

Where:

λ = component failure rate = 1 / MTBF
T = time interval (mission time or exposure time)

Time has a big impact on the probability of failure of a component. The longer the component exposure time the higher the resulting probability of failure. Exposure time refers to the length of time the component is operational or powered. As the exposure time increases, the probability of failure increases exponentially, approaching P = 1.0. Conversely, a shorter exposure results in a smaller probability of failure. The component failure rate also effects probability, the smaller the failure rate the lower the probability of failure and the larger the failure rate the higher the probability of failure.

Component exposure time can be either an advantage or disadvantage since it tends to drive the probability of failure for a component. When the hazard probability is too high, reduce the probability of failure of critical components by reducing their exposure time through a system design change.

Figure 5.9 demonstrates the effect of time on the probability of failure. In this example, a component is given a constant failure rate of 1.0×10^{-6} failure per hour (FPH) and the exposure time is varied from 1 hour to 10 million hours. The increasing probability of failure is very noticeable.

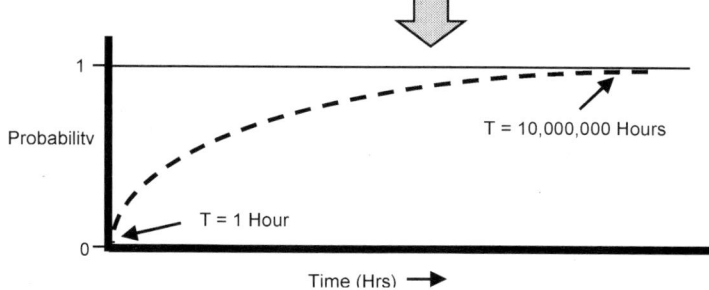

Figure 5.9 – Effect of Time on Event Probability

When performing a HA, always use the actual exposure time that components will see during system operation. Also, do not hide exposure when documenting a HA. If you are reviewing someone else's HA, always double check exposure times used in probability calculations.

5.9 Risk Acceptance Criteria

The acceptance of risk presented by identified hazards is an established process set forth in MIL-STD-882 which requires that all identified hazards be eliminated or reduced to an acceptable level of risk. In addition, the Office of the Secretary of Defense (OSD) policy clearly specifies the requirement for system safety and a formal risk acceptance process. High and Serious risk hazards can be accepted only by higher levels of authority in order to ensure that these levels of risk are not covered up and that the appropriate considerations are made in the decision. Having a person "sign off" and accept risk assures accountability and responsibility for his/her decisions. The adage of "when everyone is responsible for safety, no one is responsible or accountable" is eliminated when a person is required to sign on the bottom line that the system safety process has been followed.

The HRI Matrix concept for risk acceptance was discussed in section 5.4. The suggested mishap severity category criteria from MIL-STD-882D are shown in Table 5.1, and the suggested mishap probability category criteria are shown in Table 5.2.

Table 5.1 – Hazard/Mishap Severity Categories

Description	Cat	Mishap Definition
Catastrophic	I	Could result in death, permanent total disability, loss exceeding $1M*, or irreversible severe environmental damage that violates law or regulation.
Critical	II	Could result in permanent partial disability, injuries or occupational illness that may result in hospitalization of at least three personnel, loss exceeding $200K* but less than $1M*, or reversible environmental damage causing a violation of law or regulation.
Marginal	III	Could result in injury or occupational illness resulting in one or more lost work days(s), loss exceeding $10K* but less than $200K*, or mitigatible environmental damage without violation of law or regulation where restoration activities can be accomplished.
Negligible	IV	Could result in injury or illness not resulting in a lost work day, loss exceeding $2K* but less than $10K*, or minimal environmental damage not violating law or regulation.

Table 5.2 – Hazard/Mishap Likelihood Categories

Description	Level	Specific Individual Item	Fleet or Inventory
Frequent	A	Likely to occur often in the life of an item, with a probability of occurrence greater than 10^{-1} in that life.	Continuously experienced
Probable	B	Will occur several times in the life of an item, with a probability of occurrence less than 10^{-1} but greater than 10^{-2} in that life.	Will occur frequently
Occasional	C	Likely to occur sometime in the life of an item, with a probability of occurrence less than 10^{-2} but greater than 10^{-3} in that life.	Will occur several times
Remote	D	Unlikely but possible to occur in the life of an item, with a probability of occurrence less than 10^{-3} but greater than 10^{-6} in that life.	Unlikely, but can reasonably be expected to occur
Improbable	E	So unlikely, it can be assumed occurrence may not be experienced, with a probability of occurrence less than 10^{-6} in that life.	Unlikely to occur, but possible

Table 5.3 contains the HRI Matrix and Table 5.4 contains the risk acceptance criteria. Note that these mishap severity and probability criteria provide guidance for a wide variety of programs. The criteria in these

tables can be tailored as appropriate for a particular program. If tailoring is required, a mutual understanding of the modified terms among the stakeholders is required.

Table 5.3 – HRI Matrix

Probability		Severity			
		I Catastrophic	II Critical	III Marginal	IV Negligible
A	Frequent	1	3	7	13
B	Probable	2	5	9	16
C	Occasional	4	6	11	18
D	Remote	8	10	14	19
E	Improbable	12	15	17	20

Table 5.4 – Risk Acceptance Matrix

Risk Level	Index Values	Acceptance Criteria
High	1 - 5	Not allowed unless signed by CAE
Serious	6 - 10	Not allowed unless signed by PEO
Medium	11 - 17	Allowed; signed by PM
Low	18 - 20	Allowed; signed by PM

If tailoring of these matrices is desired, there are several aspects that can be modified, such as:

- The severity category definitions in Table 5.1 can be modified
- The probability category definitions in Table 5.2 can be modified
- The risk index cells definitions in Table 5.3 can be moved around
- The risk level index value definitions in Table 5.4 can be modified

5.10 Summary

Figure 5.10 summarizes the hazard-risk-mishap relationship. In this relationship, a hazard is a precursor to a mishap. A hazard defines a potential mishap scenario. A hazard is a specific set of conditions that truly exist within a system design. The specific set of conditions can be broken into three required elements: hazard source (HS), initiating mechanism (IM) and target-threat outcome (TTO). Hazard (mishap) risk can be computed from the three hazard elements.

Two major points to make from this figure: 1) the potential mishap cannot be understood until the TTO scenario is described by the hazard and 2) the actual risk cannot be computed until the hazard's HS and IM causal factors are thoroughly defined and understood.

The term safety is somewhat of an abstract concept. There is no measure for the amount of safety (or unsafety) in a system except for the measure of hazard risk. To determine the amount of risk (i.e., safety) presented by a system requires hazard identification, hazard risk assessment and risk reduction.

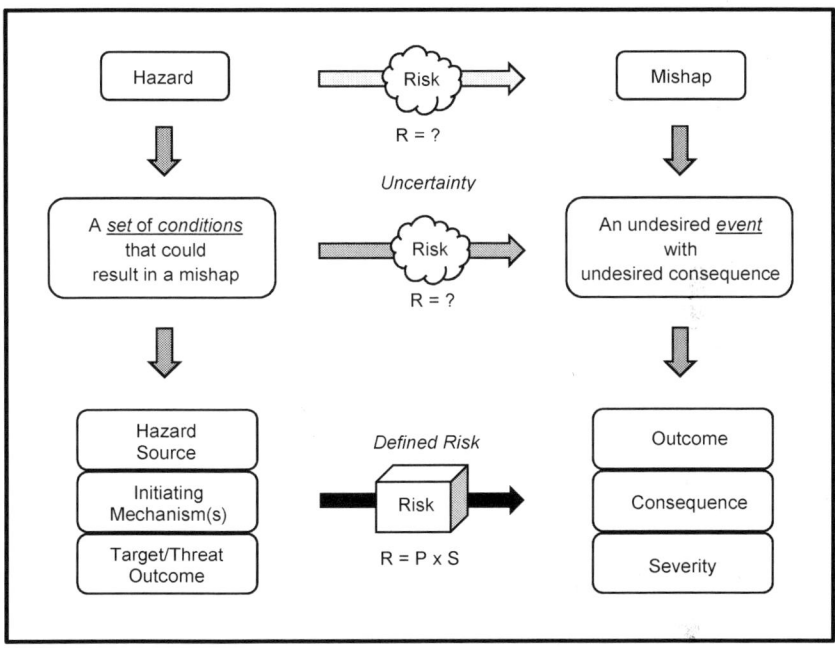

Figure 5.10 – Hazard-Risk-Mishap Relationship

Figure 5.11 summarizes how risk reduction is measured using hazard probability, severity and the HRI matrix.

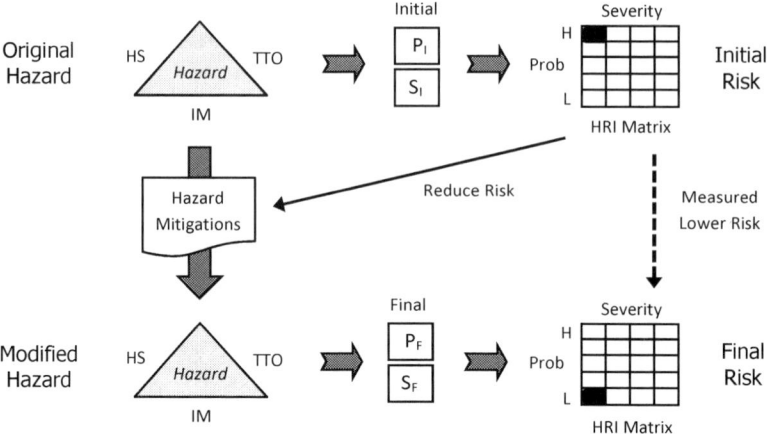

Figure 5.11 – Hazard-Risk-Mishap Relationship

Figure 5.12 demonstrates the utility and power of the SOOP concept for mitigating hazards. Mitigation via design features is the best and strongest method. It rarely fails. On the other hand, mitigation via procedures and training is weak and easily bends or fails.

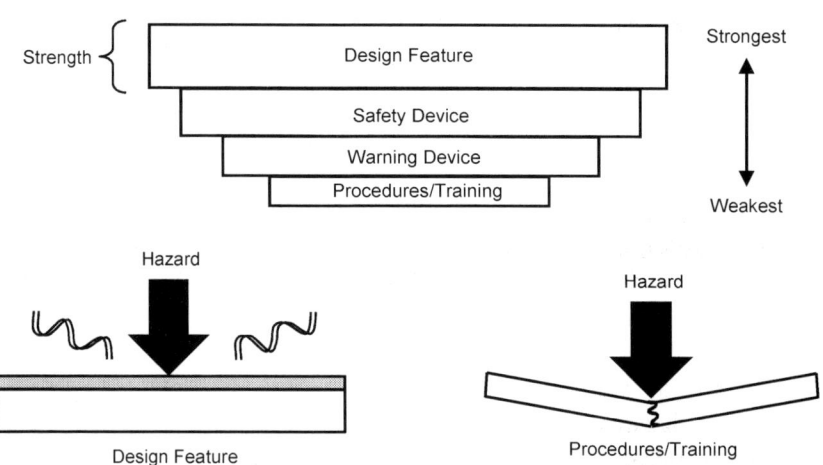

Figure 5.12 – The Strength of the SOOP Concept

CHAPTER 6

SYSTEM MISHAP MODEL

6.1 Introduction

Understanding mishaps is as important to the HA process as understanding hazards. As was shown in chapter 4, hazards and mishaps are directly linked. Identifying and categorizing mishaps is essential because it helps in understanding both hazards and mishaps; in addition it aids in providing a system mishap overview. A system mishap model (SMM) is essential in the HA process; however, it's almost a Catch-22 situation, because identifying some of the mishaps requires performing a HA. This chapter focuses on the importance of a SMM and how it can be established and utilized in the HA process. A SMM involves top level mishaps and linking them in a visual model.

6.2 Top Level Mishap (TLM)

A TLM is a generic mishap category for collecting together various hazards that share the same general outcome or type of mishap. A TLM is a common mishap outcome that can be caused by one or more hazards; its purpose is to serve as a collection point for of all the potential hazards that can result in the same outcome, but have different causal factors. TLMs provide a design safety focal point for a particular safety concern (i.e., the TLM outcome). Each contributing hazard has different initiating mechanisms or causal factors, but a common TLM outcome event. This common outcome is extracted from the hazards and used as a common TLM to unite the hazards. A TLM is essentially a *top level mishap outcome*.

During hazard analysis of a large system the number of potential hazards can became so large and diverse that the problem became one of how to easily and accurately represent the safety risks of the system design. When several different hazards can result in the same mishap, that mishap is categorized as a TLM. The TLM becomes a generic mishap category for collecting various hazards contributing to it. It is referred to as a top level mishap rather than a top level hazard because it is a collection of several different hazards, each with the same overall mishap.

Figure 6.3 illustrates the TLM concept. In this example there are five different hazards resulting in an uncontrolled aircraft fire. Each hazard has different causal factors, but a common outcome, which is an uncontrolled fire in the aircraft. This common outcome is extracted from the hazards and used as a common TLM to unite the hazards. A different hazard such as "Landing gear fails to lower" could not be placed directly under this TLM (it would fall under a different TLM).

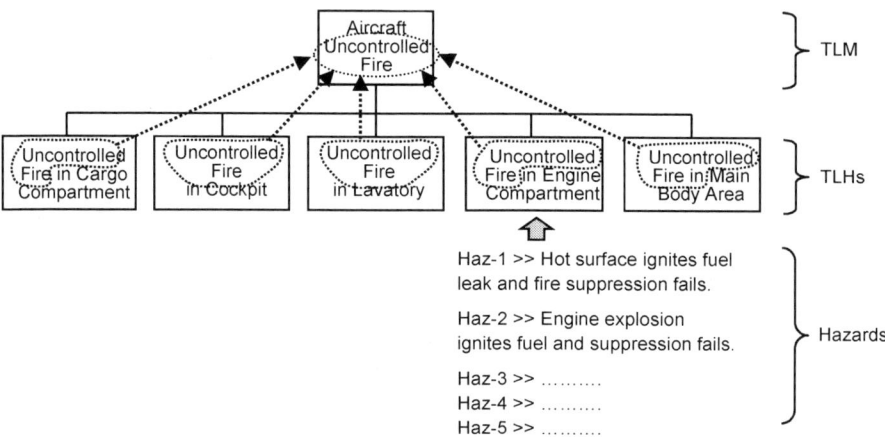

Figure 6.3 – Example TLM Derivation

Systems typically have several different TLMs, depending on the size and safety criticality of the system and the desired safety focal points. Also, different types of systems have different types of TLMs, although there may be a few similar TLMs between some system types. Some example TLMs for three different system types are shown in Table 6.1.

Table 6.1 – Example TLMs for Different System Types

Missile System	Aircraft System	Spacecraft System
• Inadvertent missile launch • Inadvertent warhead initiation • Incorrect missile target • Self-destruct fails • Electrical injuries • Mechanical injuries • RF radiation injuries • Weapon-ship fratricide	• Controlled flight into terrain • Loss of all engines • Loss of all flight controls • Loss of landing gear • Inadv thrust reverser operation • Electrical injuries • Mechanical injuries	• Loss of oxygen • System Fire • Re-entry failure • Temperature control fails • Communications fail • Electrical injuries • Mechanical injuries

"Top Level" in TLM does not necessarily imply a particular level of safety importance, but rather the common category visible at the system level (i.e., common hazards should fall within a particular TLM category). It should be noted however, that by their very nature TLMs have an implied level of safety criticality. For example, the TLM "Inadvertent missile launch" has a greater safety criticality than the TLM "Personnel injury due to electrical contact".

The value for TLMs is based on the need for system hazard clarity and focus during the HA and risk assessment process. The use of TLMs helps to focus on hazard organization. As HAs are performed many hazards are typically identified, sometimes in the thousands. With a large number of hazards it often becomes difficult to maintain hazard visibility. Sometimes hazards are inadvertently repeated; sometimes hazards are stated as causal factors rather than hazards. Since TLMs focus on the major outcome aspect of a hazard, they are useful in establishing system mishap models.

TLMs should be established early in the system safety program (SSP), generally during the Preliminary Hazard List (PHL) analysis or the Preliminary Hazard Analysis (PHA) phases. Each unique system will have its own unique set of TLMs. As TLMs are established it becomes clear where the hazard focus and hazard mitigation should be applied.

6.3 System Mishap Model

A system mishap model (SMM) is a model organizing the major potential mishaps that a system is susceptible to; it establishes and links the major mishap categories comprising a system. The purpose of a SMM is to better understand a system, its associated hazards and potential mishaps. The SMM provides HA direction and visibility. It also provides a means for summing risk at the individual mishap level. The SMM is an *a priori* (before the fact) approach to understanding hazards and mishaps.

When conducting a HA, one of the hazard identification tools is the SMM. When potential system mishap vulnerability is known, then hazards leading to that mishap category can be more easily identified. An established methodology for a SMM involves utilizing a mishap tree linking system top level mishaps. These concepts and methods are explained below.

Hazards can be recognized by focusing on known or pre-established undesired outcomes or mishaps (the TTO hazard component). This means considering and evaluating known undesired outcomes within the system. For example, a missile system has certain undesired outcomes that are known right from the conceptual stage of system development. By following these undesired outcomes backward, certain hazards can be more readily recognized. In the design of missile systems, it is understood and well accepted that inadvertent missile launch is an undesired mishap, and therefore any conditions contributing to this event would formulate a hazard, such as auto-ignition, switch failures and human error.

The SMM keeps the analyst focused on what are mishaps, what are hazards and what are causal factors. Sometimes when the HA is rushed these variable get confused and mixed. Development of the SMM is initiated prior to actual HA and finalized at HA completion. The SMM can take many forms, such as a list, spreadsheet or tree diagram.

6.4 Mishap Tree

A mishap tree is a structured diagram that organizes system mishaps into an easy to visualize and understand schema. The most effective approach for establishing a mishap tree is to utilize Top Level Mishaps (TLMs). A TLM is a generic mishap category for collecting and correlating related hazards that share the same general type of mishap event outcome. A TLM is a common mishap outcome caused by one or more hazards; its purpose is to serve as a collection point for the potential hazards that can result in the same overall TLM outcome, but have different causal factors.

TLMs help highlight and track major safety concerns and provide a design safety focal point. "Top Level" implies an inherent level of safety importance, particularly for risk visibility at the system level for a risk acceptance authority. The TLM severity will be the same as that of the highest contributing hazard's severity. Most hazards within the TLM will have the same severity level as the TLM; however, some may have a lesser severity. Figure 6.1 shows the generic concept of a mishap tree utilizing TLMs.

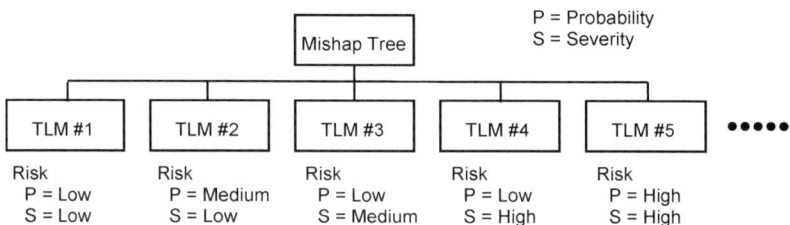

Figure 6.1 – Mishap Tree Concept

A mishap tree provides major safety focus. It identifies the potential mishap types that are safety-critical vs. those that are serious and minor in severity. Safety-critical mishap types require more attention and HA focus. Mishap tree TLMs are derived by extracting the significant and common generic outcome event portion of the contributing hazards mishap description. There are significant advantages in utilizing the TLM categorization process. It groups similar hazard, while also grouping similar risk categories. Although it's not realistic to sum the risk of all identified system hazards, it is feasible to sum the risk within TLM categories. The risk presented by all hazards cannot be summed because it is not meaningful to sum hazards with significantly different severity categories.

The mishap tree and TLMs focus on just the significant portion of the outcome described within a hazard. The TLM is restated as a simple generic outcome event when the TLM wording is shortened or slightly revised in order to make it a generic statement. TLM wording should focus on the particular safety issue of concern, such as: Inadvertent missile launch, aircraft controlled flight in terrain (CFIT), electrocution, etc.

As shown if Figure 6.2, the Mishap Tree looks strikingly similar to a fault tree. As can be seen, it shows the major mishap categories, each of which has a different likelihood and severity level.

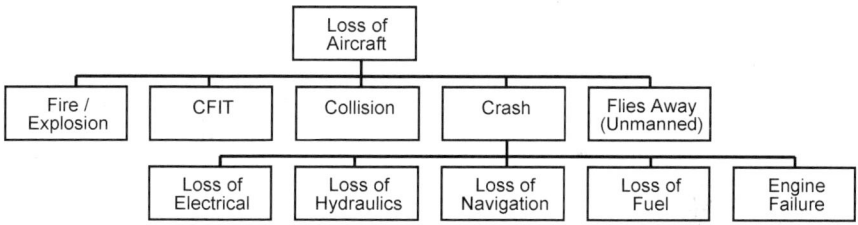

Figure 6.2 – Aircraft System Mishap Tree

6.5 Mishap Hierarchy Model

Another example of a SMM is in the form of a mishap hierarchy model (MHM). The MHM ties mishaps together in a logical and hierarchical format; it is another mishap tree diagram methodology. MHMs are graphical representations of system perturbations in the form of mishaps and mishap outcomes of concern. They are useful because they facilitate organizing thoughts and ideas into a comprehensive list of candidate hazards. An MHM resembles an FT, but it lacks explicit logic gates; it presents a hierarchical depiction of ways in which system events occur. An MHM shows the relationship of lower levels of assembly to higher levels of assembly and system function. The system hierarchies help clarify severity levels and risk levels. An MHM is recommended for the detailed identification of hazards.

Figure 6.4 shows an example MHM for an unmanned air vehicle (UAV) system.

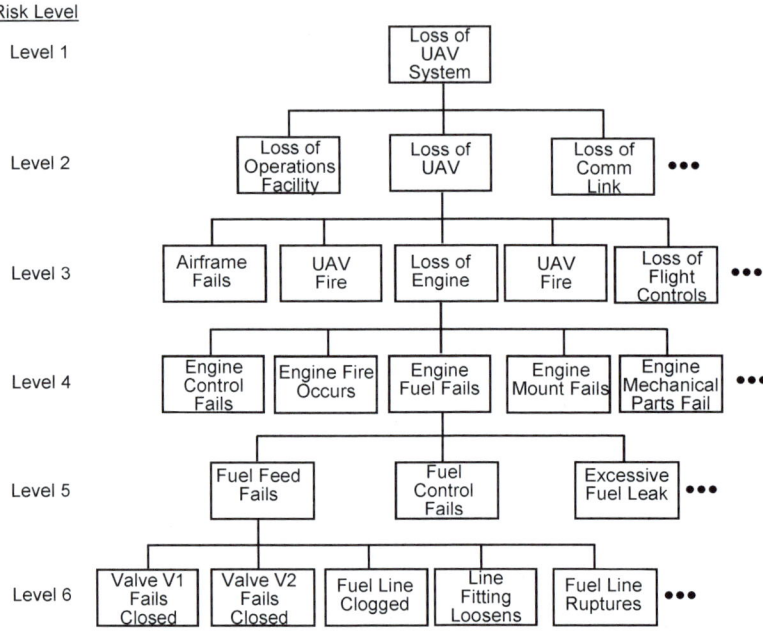

Figure 6.4 – Example MHM

6.6 Summary

In my experience, one of the most important steps in HA is the creation of a SMM. This helps to organize the entire process and aids in recognizing hazards. Hazard organization is imperative for overall system-hazard clarity and staying focused during the HA. See Appendix A for an example using mind mapping to develop the SMM.

Hazards should not be developed in an arbitrary manner, they should be developed in an organized and consistent process established by developing a SMM.

CHAPTER 7

HAZARD ANALYSIS THEORY

7.1 Introduction

Problems are typically resolved through root-cause analysis; which involves determining the causal factors to a problem and fixing them or removing them. Hazards are safety problems that must be fixed or removed. HA is more complex than typical root-cause analysis because it involves identifying both the problem and the causes. In addition, some hazards can have multiple potential causes. HA involves applying a rigorous process. There is no HA shortcut, it requires knowledge, skill and commitment. This chapter focuses on the activities required to perform a rigorous and thorough HA.

7.2 RCA vs. HA

There is a universal truism that states: nothing happens without a cause. Root Cause Analysis (RCA) is any structured methodology for identifying the basic root-cause, or causes, of a problem. RCA is based on following the thread of cause-effect relationships involved from the problem start to the final root-causal factors that lead to the occurrence of the problem. RCA involves asking "why" and "what-if" questions. Finding the root-cause is not usually achieved in the first set of questions, it typically requires asking a continuous set of questions, moving through a series of cause-effect relationships, until the answers are discovered.

RCA is typically a reactive process of identifying event causes after an event has occurred. However, when done pragmatically RCA can be used as a proactive process to forecast or predict probable causes even before they occur. And, this can be extended to HA. The HA construction process provides the logic, tools and structured methodology for asking all of the right questions needed to identify the cause-effect relationships in a system design that ultimately lead from the root-causes to the undesired event.

HA is very similar to RCA, except that in HA both the problem and the causal factors must be identified, making the process slightly more complicated. The idea in HA is to proceed logically through the system identifying system hazards and the root-causes that can cause the hazards. Figure 7.1 depicts this idea, that for each postulated hazard the *what, why how* and *when* questions must be answered in the HA. What is the undesired hazard outcome, what causes it and how is it triggered? The answer to these questions establishes the hazard and its causal factors.

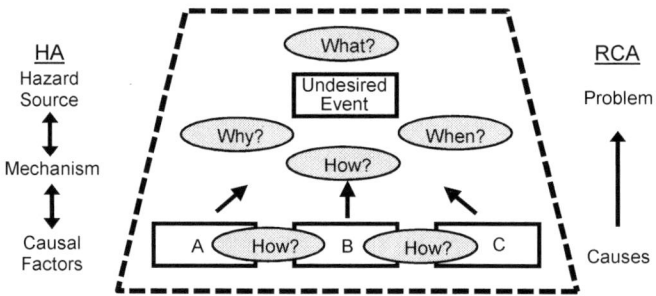

Figure 7.1 – RCA vs. HA

7.3 Hazard Analysis Tasks

Hazard identification, evaluation and control constitute the backbone of the system safety process, which is achieved through a formal HA process. Understanding and thoroughly analyzing the system is a key aspect of effective HA. The HA process is performed on a specific system design configuration, since all of the aspects of that unique design must be considered in order to identify and mitigate the system's unique hazards. If the design changes, the HA must change accordingly in order for the HA to correctly model the system.

Strictly speaking, HA could be considered as only identifying hazards. However, the HA process is tightly coupled with the system safety process, thus added steps are involved in the complete HA process. Table 8.1 lists the generic steps and specific tasks that are performed during a HA, along with representative output from each step. In general, these steps are performed regardless of the specific HA technique that is applied.

Table 7.1 – HA Tasks and Output

	Steps	Tasks	Output
1	Plan HA	• Evaluate HA requirements • Select HA technique • Establish ground rules • Establish risk criteria	- HA plan - HA ground rules - Risk criteria
2	Understand System Design	• Obtain design data • Understand design	- Design questions - System boundaries
3	Acquire HA Tools	• Identify tools • Acquire tools	- Hierarchy table - SMM - TLMs
4	Identify Hazards	• Apply HA technique • Follow ground rules • Recognize hazards • Test credibility	- Hazards - Hazard causal factors - Hazard reports
5	Validate Hazards	• Peer reviews • SSWGs	- Credible hazards - Concurrence
6	Assess Risk	• Obtain failure rate data • Obtain severity data • Calculate risk	- Hazard risk
7	Mitigate Risk	• Develop hazard defenses • Establish safety design requirements	- Safety requirements
8	Verify Mitigation	• Test result reports • Requirements verification	- Hazard closure if pass - Hazard update if failed
9	Accept Risk	• Package hazards • Obtain approvals	- Risk acceptance letter - Signatures
10	Track Hazards	• Record hazards • Track hazards/risks • Close hazards	- Hazard database - Safety case - Lessons learned

7.3.1 Plan HA

Properly planning the HA process for a program is one of the most important and critical steps in the HA process. Performing a good HA is not a simple or trivial task; it requires foresight, planning, methodology, organization and a total systems viewpoint in order to achieve uniformity, consistency and full system coverage. It also requires someone with experience and skill in safety and HA. Quite often analysts just jump into a system and immediately start identifying what they think are hazards, without considering the system architecture and the overall risk objectives and relationships for the HA. This approach often leads to confusion, overlap, gaps and hazard-risk mismatches.

The HA plan should define and document the many aspects and parameters of a HA, such as:

- Hazard and HA definitions
- How to write hazards for the HA (context)
- HA guidelines
- HA scope, ground rules and schedule
- Risk rating criteria
- Risk acceptance criteria
- HA techniques to be used
- System level at which hazards-risk should be written
- The program risk management process
- HA tools to be used

Since there are so many complexities involved with hazards and HA, a *system hazard architecture* should be developed as part of the planning process. This architecture should define TLMs, SMM, system hierarchy, and where risk should be tabulated in the system hierarchy. Hazards should not be developed in an arbitrary manner, they should be developed in an organized and consistent process.

7.3.2 Understand System Design

Each HA is performed on a specific system design configuration, in order to make that specific design safe. If the system design and method of operation is not thoroughly understood, the HA will likely not be effective and may not even be entirely correct. This step involves obtaining, reviewing and understanding system drawings, schematics and design documents. It also involves attending design meetings and reviews during the system development.

7.3.3 Acquire HA Tools

Certain tools are typically utilized during the performance of a HA. These tools are typically identified during the planning process and acquired prior to initiating the HA. Many of the tools used during HA are described in chapter 8. HA tools aid in understanding system design and operation, and they aid in recognizing and identifying hazards.

7.3.4 Identify Hazards (or Hazard Recognition)

Identifying, or recognizing, hazards is the key to HA and probably the most difficult and involved task of all. Hazard recognition is not a trivial process, and it is made even more difficult by the many HA complexities involved (see chapter 15). HA requires a thorough understanding of what exactly comprises as hazard (see chapter 4), as well as a system hazard roadmap (see chapter 6). An agreed upon definition for a hazard should be included in the HA plan. In many respects, HA is a skill and art that requires a knowledgeable and experienced analyst. Section 7.4 provides detailed information of how to specifically recognize hazards during the hazard identification phase.

7.3.5 Validate Hazards ~Threat~

Hazard validation involves reviewing postulated hazards and obtaining concurrence that the hazards are credible and real. Typically, hazards are reviewed by a separate group of individuals after they have been established by the safety group. A peer review of hazards is conducted by technical areas experts, or subject matter experts (SMEs), in the technical areas covered by the HA. For example, after completing the initial HA of an aircraft system, SMEs on aircraft engines would review all engine related hazards for concurrence. The SMEs might recommend rewording hazards, deleting hazards or adding hazards. System Safety Working Groups (SSWGs) are often held to review postulated hazards in the same manner. A SSWG can be comprised of any stakeholders in the system design, development and operation.

7.3.6 Assess Risk

Mishap risk is the safety metric characterizing the amount of danger presented by a hazard. Risk is the likelihood that a hazard will result in an actual mishap, multiplied by the mishap severity or amount of expected loss to be incurred from that mishap. Risk provides a predictive measure that system safety uses to rate the safety significance of a hazard and the amount of improvement provided by hazard mitigation.

The risk management process used by system safety typically utilizes an HRI matrix to identify and assess the amount of safety risk presented by a hazard. The HRI matrix is a risk ranking scheme used in system safety to distinguish lower risk hazards from higher risk hazards. The HRI matrix ranks the safety criticality of each hazard through the use of a risk index and a risk level. Once an HRI matrix has been established, hazard risk can be evaluated according to the HRI matrix definitions. The HRI matrix is a tool for evaluating, judging and accepting or rejecting risk.

7.3.7 Mitigate Risk

The risk management process is a key element in developing a system that presents acceptable mishap risk resulting from residual hazards. An effective hazard-mishap risk management process must be formulated for each program. The program HRI matrix is used to rank the safety criticality of each hazard through the use of a HRI and a risk level. The HRI determines if the hazard risk must be mitigated, and the hazard risk level establishes who accepts the risk.

When the risk presented by a hazard is determined to be unacceptable, it must be mitigated. Mitigation is typically done through the implementation of a design safety feature (DSF), such as a redundant component, fail-safe mechanism, interlock, etc. System Safety Requirements (SSRs) are the vehicle that establishes the DSFs for the system design. By implementing the DSFs contained in the SSRs, the mishap risk is reduced to an acceptable level. But, it does not end there; the SSRs must be verified and validated (V&V) to ensure they are implemented and effectively eliminate the hazard or reduce it to an acceptable level of risk. Mitigation success is verified through appropriate analysis, testing, demonstration or inspection. The use of a traceability matrix provides a trace from the hazard to the mitigating SSRs, and then to the test requirements, to the test results.

SSRs are a combination of generic and derived safety requirements. Generic SSRs typically come from existing safety documents and standards, such as special safety requirements for laser safety. These are SSRs established from historical hazards from other systems. Derived SSRs are established for the mitigation of hazards identified for the system under development. SSRs are included in the system specification.

7.3.8 Verify Mitigation

The purpose of this step is to ensure that the planned hazard mitigation is carried out. It is also to validate the success of the implemented risk mitigation measures. System safety must take action if the mitigation plans are not being followed. System safety must also perform follow-up analysis and data reviews to ensure that the implemented mitigations are effective. This step involves ensuring that the safety design requirements are documented, tested and that they successfully pass testing.

7.3.9 Accept Risk

DoD and OSD policy mandates that the risk for all identified hazards must be accepted by the appropriate risk acceptance authority. The HRI matrix plots mishap likelihood along the y-axis and mishap severity along the x-axis, with each axis typically divided into 3 to 5 discrete columns and rows. Each combination of frequency and severity intersect at a particular matrix cell, referred to as a HRI, because the cells are numbered by severity criticality ranking. The combination of severity indexes for risk levels of High, Serious, Medium and Low. Low and Medium risk hazards are generally considered acceptable at the program level. Serious and High risk hazards are considered as unacceptable, unless accepted by a high ranking responsible person that has made the decision based on specific criteria. Evidence of completion is a documented risk acceptance letters that are officially signed by the appropriate risk acceptance authority.

7.3.10 Track Hazards

Hazard tracking is the process of systematically recording all identified hazards and the data associated with these hazards as they progress through the lifecycle of a hazard: identification, assessment, mitigation, verification, acceptance and closure. It is typically achieved through the use of a formal hazard tracking system (HTS). Hazard tracking is a basic required element of an effective system safety program (SSP), essential to knowing the status of all hazards at any point in time, in order to none go unresolved. It also provides a record of all activity and information associated with each hazard. It is usually referred to as closed-loop hazard tracking because it is similar to a control system with a feedback loop that is active and allowing for changes and updates, and the reiterative performance of some risk management tasks. As changes are incorporated into the system, the HTS is updated to reflect changes in hazards and residual risk. The HTS is a living document which needs to be revisited periodically throughout the system lifecycle to determine if accepted hazards can be eliminated or further reduced through new technology or process changes.

The HTS retains the open/closed status of all hazards. The status of a hazard is "Open" until it has been verified that the appropriate safety mitigation methods have been established to mitigate the risk; it then moves to "Monitor" status. When the mitigation methods have been implemented and proven successful through testing, and the risk has been accepted by the appropriate risk acceptance authority, the hazard's status is changed to "Closed". It is important to remember that a closed hazard does not mean no risk exists; it is an acknowledgement and acceptance of a system hazards and its residual mishap risk. Chapter 8 discusses the HTS tool in more detail.

7.4 Hazard Recognition

7.4.1 Hazard Recognition Introduction

Hazard identification, or hazard recognition, is not a trivial process; it requires a thorough understanding of what exactly comprises as hazard, as well as a grounded understanding of the system under investigation. In many respects, HA is a skill and art that requires a knowledgeable and experienced safety analyst. Hazard identification must be taken seriously because of the safety-related repercussion involved.

Hazard recognition is the cognitive process of visualizing a hazard from an assorted package of design information and hazard knowledge. In order to recognize or identify hazards four things are necessary:

1) An understanding of hazard theory
2) A hazard analyses technique to provide a consistent and methodical process
3) An understanding of hazard recognition methods
4) An understanding of the system design and operation
5) An understanding of system hazard organization

HA involves the recognition and identification of system hazards and causal factors using a systematic and methodical approach, which is typically achieved through the application of several different system safety hazard analysis techniques. The HA plan should describe the combination of analyses that will be used to identify hazards for the particular project. HA should evaluate hardware, software, firmware, procedures and HSI. HA should consider all lifecycle phases of the system, as well as environmental factors. Consideration should also be given to proposed design changes, software trouble reports and technology upgrades, just to name a few.

7.4.2 Hazard Recognition – System Perspectives

When performing a HA the system must be viewed from different perspectives (discussed in chapter 3), each of which provides a different viewpoint and understanding of the system. These system viewpoints include:

1) Physical – this view involves the various architectural views that depict what the system contains and how it is constructed. This view establishes subsystems, assemblies, components and the overall system equipment hierarchy.

2) Functional – this view evaluates what the system must do in order to produce the required system behavior, broken down into functions with input, output and transformation rules.

3) Operational – this view defines how the user will view and operate the system, including instructions, tasks, conditions, parameters and limitations.

4) Software – this view looks at the system software, which is difficult to fully grasp and is somewhat abstract.

5) Environment – this view looks at the various environments that the system will encounter (internal and external). This includes natural environments (e.g., weather, tornadoes) as well as system environments (e.g., heat, EMI).

6) Human – this view looks more closely at human performance in the system and the effect of potential human errors. It also includes user interfaces with the system and their potential impact on the user.

7) Organizational – this looks at the organizational and management causal factors affecting a hazard.

It is important that HA consider and evaluate a system from each of these perspectives in order to ensure complete safety coverage of the system. This is why HA must consider and focus on system functions, system operations, and system components, including hazardous energy sources. This also explains why more than one type of HA must be applied in order to identify all hazards, because one HA type alone does not typically provide sufficient hazard identification coverage.

7.4.3 Hazard Recognition – Failure Perspectives

All of the basic system components and characteristics must be understood in order to perform a complete and thorough hazard analysis. Examples of typical safety considerations for various system elements include:

- Hardware – failure modes, hazardous energy sources

- Software – design errors, design incompatibilities
- Personnel – human error, human injury, human control interfaces
- Environment – weather, external equipment (e.g., radiation, chemicals)
- Procedures – instructions, tasks, warning notes
- Interfaces – erroneous input/output, unexpected complexities
- Functions – fails to perform, performs erroneously, performs inadvertently
- Facilities – construction factors, storage factors, transportation factors

A HA must adequately consider and address each of these system attributes and their interrelationships in order to ensure that all possible hazards are identified. For example, it is possible for different operational phases to have different safety impacts; different functions performed during a phase could have a direct impact on subsequent phases. During certain phases safety barriers or interlocks are often removed, making the system more susceptible to the occurrence of a hazard, e.g., at one point in the operational mission of a missile, the missile is powered and armed. This means that fewer potential failures are now necessary in order for a mishap to occur and there are fewer safeguards activated in place to prevent hazards from occurring.

Hazards are typically actuated as the result of any of the following system initiating mechanism factors, or combinations thereof:

- Hardware failures (aging, wear, random failure)
- Software errors/flaws (functional failure) – Hack
- Human errors (performance, decisions, judgment)
- Design errors/flaws (interface errors, sneak paths)
- External environmental factors (EMI, lightning)
- Maintenance flaws/errors (resulting in failures)
- Manufacturing flaws/errors (resulting in failures)
- Particular risk events (events occurring outside of a subsystem, such as a flood, fire, etc.)

In addition to typical hazard sources, failures, flaws and errors can be also created or perpetuated by the following organizational factors:

- Poor design/development/manufacturing processes

Hazard Analysis Primer

- Organizational errors (performance, decisions, safety as a core value)
- Lack of safety culture in overall organization/company
- Safety organizational level of competence

It is important to note that hazards can exist and occur without the presence of a hardware failure mode or software error. Subtle design flaws can produce hazards, such as a sneak path in an electrical circuit. When performing an HA, standard considerations for identifying hazards include:

- Safety-related functions (SRFs) and safety-critical functions (SCFs) (hardware/software)
- Hazard Sources (e.g., energy sources, SCFs, environments, particular risk sources)
- Hazardous Assets
- Hardware failures (component failure modes)
- Sneak paths *Threat Don't exist.*
- Safety-related (SR) and safety-critical (SC) human tasks

7.4.4 Key Hazard Recognition Factors

Some key recognition factors that can help the safety analyst visualize hazards, include:

1) Hazard sources
2) Hazard checklists
3) Lessons learned
4) Safety criteria
5) Key failure state questions
6) Good design practices
7) TLMs
8) SCFs
9) The SMM and its overall layout
10) Hazard Triangle components

Hazard Sources
Identify the hazard sources in the system. Each hazard source will be responsible for one or more hazards. Use the system hierarch table, list of equipment and hazard checklists to identify hazard sources in the system.

Hazard Checklists — Risk Register
Use hazard checklists to identify hazards in the system. The checklists are mental reminders of items, functions and procedures that are hazardous. A good safety analyst should have many different hazard checklists at hand. Refer to Chapter 11 for some example checklists.

Lessons Learned
Utilizing past knowledge from experience and lessons learned. Review lessons learned documents and databases. Mishap and hazard information from a previously developed system that is similar in nature or design to the system currently under analysis will aid in the hazard recognition process. For example, by reviewing the mishaps of an earlier spacecraft system, it might be discovered that the use of a particular seal design on a specific part of the spacecraft resulted in several mishaps. Seal design on this part of the spacecraft should then be recognized as a potential hazardous area requiring special focus in new spacecraft design.

Safety Criteria
Review general design safety criteria, precepts and principles applicable to the system under investigation. Use the criteria to identify specific hazards in the system design. By considering the reasoning and logic behind specific safety criteria, precepts and principles some types of hazards can be more easily recognized. For example, there is a good safety reason for the following safety criteria "Do not supply primary and redundant system circuits with power from the same bus or circuit". This safety criterion is a clue that aids in recognizing hazards in systems involving redundancy. A hazard analysis should look for such hazards in a new design.

Key Failure State Questions
Use key failure state questions to identify hazards. This is a method involving a set of clue questions that must be answered, each of which can trigger the recognition of a hazard. The key states are potential states or ways the subsystem could fail, or operate incorrectly and thereby result in creating a hazard. For example, when evaluating each subsystem, answering the question "what happens when the subsystem fails to operate?" may lead to the recognition of a hazard. For everything in the

system (item, function, human, task, etc.) ask what happens if the item does one of the following:

- Fails to operate
- Operates incorrectly/erroneously
- Operates prematurely/inadvertently
- Operates out of sequence
- Operates in degraded mode
- Unable to stop operation as intended
- Sends/receives/displays erroneous data
- Causes operator confusion

Good Design Practices
Analyze known and recommended good design practices. If these practices are not in effect, then there may be hazards associated with these practices. By reverse engineering good design practices found in various design documents and standards, hazardous design situations can be identified. For example, a good design practice for interrupting DC circuits is to place the switch or circuit breaker on the power branch, rather than the ground branch of the circuit. This prevents inadvertent operation should a fault to ground occur.

TLMs and SCFs
Review identified TLMs and SCFs to determine if any additional hazards have been overlooked in these categories.

SMM
Review the SMM to determine if all hazards have been identified for each of the undesired outcomes and TLMs.

Hazard Triangle
 The Hazard Triangle concept provides a hazard recognition resource by evaluating individually each of the three hazard component categories from a systems context. For example, identify and evaluate all of the HS components in the unique system design as the first step. Next, evaluate all of the potential system failure modes to determine which failures are critical IMs for hazards. Then, identify and evaluate all potential targets and the possible threats to them.
 When considering the HS component of the Hazard Triangle, focus on hazardous element checklists. The HS component involves items known to be a hazardous source, such as explosives, fuel, batteries, electricity,

acceleration, chemicals, and the like. This means that generic hazard source checklists can be utilized to help recognize hazards within the unique system design. If a component in the system being analyzed is on one of the hazard source checklists, then this is a direct pointer to potential hazards that may be in the system. Speculate all of the possible ways that the Hazardous Element can be hazardous within the system design. There are many different types of hazard source checklists, such as those for energy sources, hazardous operations, chemicals, and so forth.

Hazards can be recognized by focusing on known hazard triggering mechanisms (the IM hazard component). For example, in the design of aircraft it is common knowledge that fuel ignition sources and fuel leakage sources are initiating mechanisms for fire/explosion hazards. Therefore, hazard recognition would benefit from detailed review of the design for ignition sources and leakage sources when fuel is involved. Component failure modes and human error are common triggering modes for hazards.

Hazards can be recognized by focusing on known or pre-established undesired outcomes or mishaps; review the potential targets and threats. This means considering and evaluating known undesired outcomes within the system. For example, a missile system has certain undesired outcomes that are known right from the conceptual stage of system development. By following these undesired outcomes backward, certain hazards can be more readily recognized. In the design of missile systems, it is well accepted that inadvertent missile launch is an undesired mishap, and therefore any conditions contributing to this event would formulate a hazard, such as auto-ignition, switch failures and human error.

7.4.5 Hazard Recognition Basics

When identifying hazards, a key factor to keep in mind is the understanding of why hazards exist in the first place. Knowing why and how they exist helps in their identification. Hazards exist primarily for three reasons: 1) hazardous assets are used in the system, 2) design miscalculations and errors occur in the design process, and 3) failures, human errors and environments impact the design during system operation.

Hazards are created because of the need for hazardous sources in the system, or they must interface with hazard sources, coupled with the fact that eventually everything fails, and these failures can unleash the undesired effects of the hazard source. Hazards also exist due to the need for safety-critical system functions, coupled with the potential for failures and human error within these safety-critical functions. Hazard creation can be summarized by the following factors, which can occur singularly or in combinations:

- The use of hazardous system elements (e.g., fuel, explosives, electricity, velocity, stored energy).
- The system interfaces with hazard sources (e.g., fuel, electricity).
- Operation in hazardous environments (e.g., flood zones, ice, heat).
- The need for hazardous functions (e.g., aircraft fueling, welding).
- The use of safety-critical functions (e.g., flight control, arming).
- The inclusion of (unknown) design flaws, errors and sneak paths.
- The potential for hardware wear, aging and failure.
- Inadequacy in designing to tolerate critical failures.

7.4.6 Hazard Recognition Sources

Table 7.2 lists various sources that can be used to recognize and identify hazards.

Table 7.2 – Hazard Recognition Methods

	Hazard Recognition Subject Matter	Recognition Thoughts
1	Energy sources	Identify the various hazards that can spawn from energy sources used in the system. Consider different scenarios, such as if the energy source fails to function, functions inadvertently, becomes uncontained, accidently contacted, etc.
2	Critical system functions	Determine which system functions are needed for safe system operation. Consider different scenarios, such as if the function fails, functions inadvertently, functions out of tolerance, etc.
3	Redundant items	Identify the redundant subsystems and assemblies in the system. Determine which are safety-critical. Consider different scenarios, such as if the function fails, functions inadvertently, functions out of tolerance, etc.
4	Expected environments	Identify all of the potential environments that could be possibly encountered. Identify systems/equipment that may be sensitive to certain environments.
5	System hierarchy table	Evaluate every item on the hierarchy table. Identify items that are hazardous assets or hazardous if they malfunction.
6	Hazardous assets	Determine which equipment in the system hierarchy table are hazardous in some way, and postulate the hazards that might be possible. For example, aircraft flight controls are hazardous if they fail.
7	Existing design safety features	Look at the design of safety features claimed in the system design and evaluate the impact if they should fail or malfunction.
8	TLMs / SMM	While looking at the TLMs and the SMM; consider the items in the hierarchy table and determine if any can contribute a hazard to one of the categories.

9	Hazardous subsystems	Determine which subsystems in the system hierarchy table are hazardous in some way, and postulate the hazards that might be possible. For example, aircraft flight controls are hazardous if they fail.
10	Critical software functions	Determine which software functions are needed for safe system operation. Consider different scenarios, such as if the function fails, functions inadvertently, functions out of tolerance, etc.
11	System dependencies	Identify equipment that can cause hazards if they are caused to fail simultaneously by common cause faults.
12	Hazard checklists	Review hazard checklists to see if they trigger ideas for any hazards in the system design.
13	Lessons Learned	Review lessons learned and mishap reports to see if they trigger ideas for any hazards in the system design.
14	Design Requirements	Review design requirements to identify hazardous situations created by the requirements and/or missing requirements affecting safety.
15	Hazard Organization	Organizing hazards below TLMs and TLHs helps in visualizing and writing hazards. Be sure to establish TLMs and TLHs during the process.

7.5 Describing the Identified Hazard

A very important aspect of HA that is often overlooked has to do with describing an identified hazard. A good hazard description depends upon the proper context; it describes a scenario that includes the hazard elements of HS, IM and TTO. It should include only the basic hazard causal factors (HCFs) needed to cause the hazard and nothing more. If more is possible, then that is usually material for another similar hazard.

A hazard is very similar to a mini-FTA. A hazard requires a HS, IMs and TTO. A FTA begins with an undesired event (outcome) and establishes the contributing causal factors down the FT structure (HS and IMs). A FT contains the specific causal factors leading to the occurrence of an undesired event; the undesired event is the outcome and the causal factors are the IMs. Where a FTA usually has many unique cut sets that cause the undesired event, a hazard is a single cut set. A hazard is more like a cut set in a FTA, rather than a large FTA. Many different hazards leading to the same undesired event can be linked together by the FT structure. A hazard is a min cut set of a larger group of hazards; they all fall under the same general TLM. For example, Inadvertent Missile Launch can be caused by many different potential failures, each of which is a separate hazard. The FT combines them together for a single probability of occurrence for the sum of the individual hazards.

Sometimes analysts try to put too much information into a hazard, as well as too little. With too little information the hazard cannot be fully comprehended or the risk calculated. Too little information produces pseudo hazards, such as "aircraft collision". With too much information

Hazard Analysis Primer

there are usually multiple causal ORed factors included, which should be divided into separate hazards. For example, "inadvertent missile launch occurs due to switch fails closed OR human error closes switch". Note: the OR operator refers to the Boolean OR operator used in probabilistic mathematics; as opposed to the AND operator.

Figure 7.2 shows a fault tree (FT) for an undesired event of "Operator Amputates Finger on Table Saw". This FT has two cut sets:

1) A • B   (A and B)
2) A • C   (A and C)

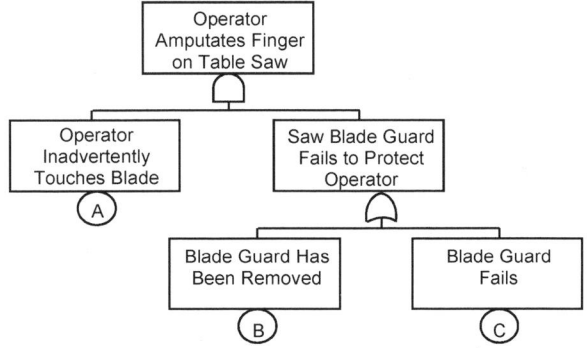

Figure 7.2 – FTA of Table Saw Hazard-Mishap

If the events in this FT were written as a hazard, it would be tempting to write the hazard as:

"Operator amputates finger while using table saw because the operator inadvertently touches the saw blade and the blade guard is removed or it fails".

The FT, however, actually represents two hazards since there are two unique cut sets. The two hazards would be written as follows:

Hazard-1: "Operator amputates finger while using table saw because the operator inadvertently touches the saw blade and the blade guard is removed".

Hazard-2: "Operator amputates finger while using table saw because the operator inadvertently touches the saw blade and the blade guard fails".

It may seem like a small point that two hazards should be written rather than one hazard; however, there is sound reasoning for this. When the HCFs are ORed (summed) together, the probability is greater than for each individual hazard. This confuses the risk decisions that must be made. For example, is it more important to prevent guard removal of mitigate guard failure? The relative risk for each hazard must be calculated for comparison and decision making.

Note that not combining ORed hazard causal factors under one hazard is a general rule; however, there are occasionally exceptions. For example, sometimes it is necessary to capture risk at a higher system level and then the hazard causal factors are combined.

Write hazards to ensure they are characterized in complete system context:

- Describe the Hazard Source (HS), Initiating Mechanism (IM) and Target/Threat Outcome (TTO)
- Do not abbreviate or assume readers understands program-special lingo and acronyms
- Describe the hazard scenario in context (e.g., "fuel" is not a hazard, but, "fuel leak and an ignition source leading to fire and system loss" is a hazard)
- Make sure the hazard context includes the specific hazard causal factors and the specific hazard-mishap effects
- Write the hazard in a complete sentence

7.6 Summary

One of the most important factors in HA is correctly describing an identified hazard. Safety analyses and risk assessments are only as good as the hazard descriptions they are based on.

Hazards occur in a system for one simple reason – the system must utilize, or interface with, a hazard source (hazardous asset) in order to achieve its intended goals. Can a hazard exist without a HS? Going back to the HS-IM-TTO concept and the hazard triangle, the answer is no. However, the HS can include a diverse set of factors. A HS can be an energy source, a safety-critical function, an adverse environment, etc. A mishap occurs for two reasons – a hazard exists in the system design and the hazard is not eliminated or is not appropriately mitigated.

CHAPTER 8

HAZARD ANALYSIS STEPS

8.1 Introduction

The theory and concepts leading up to the HA process have been discussed in the previous chapters. This chapter focuses on the specific basic steps to apply in the hazard HA, which are relatively simple, straight forward and generally apply to all HA techniques.

8.2 The Ten Essentials of Hazard Analysis

The ten essentials of HA is a list that can be used as a reminder of the basic essential items to have in place when performing a HA. These items are listed in Table 8.1.

Table 8.1 – Ten Essentials of HA

	Essential Items
1	System design and operation data
2	System design specifications and requirements
3	HA plan (scope, guidelines, etc.)
4	Hazard theory understanding
5	Understanding how to write a hazard description
6	HA tools (system hierarchy table, checklists, etc.)
7	System Mishap Model (SMM) for the system
8	Risk rating criteria (for hazard rating and acceptance)
9	Hazard tracking system
10	Knowledgeable in the HA technique selected

8.3 General HA Steps

Regardless of the specific HA technique (methodology) that is applied, there are certain steps that should always be performed. Table 8.2 lists the general HA steps to serve as a reminder of what is essential and necessary when conducting a HA.

Table 8.2 – Primary HA Steps

	Step	Purpose
1	Develop a list of system equipment (hardware and software)	To identify safety-critical items and to ensure HA covers all equipment.
2	Develop a list of system functions	To understand system and to identify safety-critical functions.
3	Develop a system functional diagram	To understand system operation, redundancies and dependencies.
4	Develop a system hierarchy table	To identify and keep track of the equipment and subsystems to be analyzed.
5	List overall safety concerns for system	To identify obvious undesired outcomes from the system; this will help establish TLMs.
6	List design safety features existing in system design	To identify design strengths and weaknesses.
7	Identify general hazards and mishap categories	To establish TLM outcomes and initial hazards.
8	Establish SMM	To develop a logical map of hazards, outcomes, and interrelationships. To organize hazards for clarity and consistency.
9	Identify hazards using hazard recognition aids and tools	To identify hazards using the system tools and recognition tools; the SMM is the overall guide.
10	Check hazards for wording, consistency and correctness	Review and peer review hazards to ensure they correct.

8.4 General HA Perspectives

Table 8.3 contains the various system views that should be covered by a HA. It is important to ensure that each view is covered by the HA and to provide some sort of evidence verifying complete coverage.

Table 8.3 – Checklist for System Perspectives

System Perspective	Safety Considerations
Physical	This view involves the various architectural views that depict what the system contains and how it is constructed. It establishes subsystems, assemblies, components and the overall system equipment hierarchy. The HA should identify these physical elements and show that each was covered.
Functional	This view involves the functions that the system must perform in order to operate. All functions should be identified and evaluated by the HA. Safety-critical functions will be identified at this stage.

Hazard Analysis Primer

Operational	This view involves how the system will operate and how the user interfaces with the system, including instructions, tasks, conditions, parameters and limitations.
Software	This view the system software. Primarily, software functions are identified and evaluated by the HA. Safety-critical software functions will be identified at this stage. Functional software hazards must be also carried into the hardware and human aspects involved.
Environment	This view considers the various environments that the system will encounter (internal and external). This includes natural environments (e.g., weather, tornadoes) as well as system environments (e.g., heat, EMI).
Human	This view considers human performance in the system and the effect of potential human errors. It also includes user interfaces with the system and their potential impact on the user. Human error should be looked at from all perspectives, including how the system design can force the user to commit an error.
Organizational	This considers the organizational and management aspects affecting a hazard.

8.5 General HA Checklist

The following is a general HA checklist that is provided to serve as a reminder of what is essential and necessary when conducting a HA:

1) Design requirements have been read and understood.
2) Design and operational data have been read and understood.
3) Operational phases have been identified.
4) A system hierarchy table has been developed to ensure complete system coverage.
5) The various system have been identified and evaluated:
   - Energy sources (hazardous assets)
   - Hardware components and functions
   - Critical system functions
   - Hunan interfaces with system
   - Critical human tasks
   - Redundant items
   - Expected environments
6) The appropriate HA scope, boundaries and limitations have been identified and established.
7) Existing design safety features have been identified and documented.
8) Industry hazard checklists have been compiled and reviewed.
9) The hazard titles and descriptions are unique and there are no duplications or overlaps.

10) The severity component of risk is consistent with severity at the worst credible level.
11) The probability component of risk is consistent with experience and/or historical data.
12) The correct mission exposure time has been established for risk probability calculations.
13) Hazard descriptions fulfill the Hazard Triangle and include the hazard elements of HS, IM and TTO.
14) Hazard descriptions include full context of the hazard; rewrite as necessary until the hazard description is thorough and correct.
15) All postulated hazards have been recorded and documented, even when they have been determined to be non-credible.
16) The established hazard countermeasures are adequate and do not impair operations or performance.
17) Established hazard countermeasures do not create new hazards.

8.6 Summary

Systems are imperfect mechanisms developed to achieve an intended capability; unfortunately they often have unintended side effects when misused, abused or failures occur. When an item is evaluated for safety and reliability it must be evaluated by itself and also as it functions within a system environment. When an item is in a system it is exposed to a whole new set of dynamics and environments that it does not see separately from the system. These new dynamics present a new set of hazards and risk factors.

The key to hazard identification is *hazard organization*. This is the reason a SMM has been continuously underlined. Too many HAs go bad because the hazards (and the HA) are not organized properly. Hazard organization provides the analyst, and HA reviewers, with an overview and a map of the overall TLMs, hazard considerations, hazard paths and hazards.

CHAPTER 9

HAZARD ANALYSIS TOOLS

9.1 Introduction

Hazard identification, evaluation and control constitute the backbone of the system safety process. The only sure way to thoroughly identify hazards is through the application of rigorous, structured and formal HA techniques. In achieving this goal there are many different tools available to the analyst for hazard recognition, recording, evaluation and tracking. This chapter focuses on some of the major tools used in HA.

9.2 HA Tool Categories

The safety analyst has many tools available to assist in the HA process. Tools in the HA toolbox fall into the following categories:

- Hazard identification techniques
- HA support tools
- Hazard recognition aids
- Historical data
- Hazard tracking tools

Hazard Identification Techniques

A formal HA technique is necessary to provide analysis rigor and also to provide analysis evidence. Typically, multiple HA methodologies are applied to a system to ensure complete system coverage and complete hazard identification. This is necessary to provide complete safety coverage of the system viewpoints mentioned earlier: physical, functional, operational, software, environment and human (see chapter 7). The number and type of HAs to be performed should be planned in advance and tailored to the specific system and the system requirements involved, depending upon various factors, such as: system size, system complexity, safety criticality of the system, etc.

HA Support Tools

HA is a complex and thoughtful process, which can be simplified with the help of several design development tools that are usually available on a development program. The tools most useful in HA are typically produced by the systems engineering and reliability engineering disciplines.

A system can be very large, comprising hundreds of subsystems, thousands of assemblies and millions of components. The systems engineering process utilizes many different tools during systems development that aid in the understanding of large and complex systems. Some systems engineering tools that greatly aid the system safety analyst are:

- Simplified System Diagrams
- Functional Block Diagrams
- Master Equipment List (MEL)

Some reliability engineering tools that greatly aid the system safety analyst include the following:

- Reliability Block Diagrams (RBDs)
- Failure Mode and Effects Analysis (FMEA)

Hazard Recognition Aids

Hazard recognition is the cognitive process of visualizing a hazard from an assorted array of design information. Hazard recognition and identification is not always an easy process. Hazards are like ubiquitous and elusive creatures that must be hunted and captured, where the hunt is akin to the HA process. There are, however, some aids which assist the analyst in the process, such as:

- Hazard Checklists
- Mishap Checklists
- Hazard Recognition Keywords
- Historical Data

Some example keywords that can be used to assist in identifying hazards include the following:

- Fails to function (operate)
- Functions incorrectly/erroneously
- Functions prematurely/inadvertently
- Functions erroneously (out of sequence, degraded mode)

- Unable to stop function as intended
- Sends/receives/displays erroneous data
- Causes operator confusion

Historical Data

Another method for recognizing hazards is through the use of past knowledge from experience and lessons learned. Safety lessons learned from a previously developed system that is similar in nature or design to the system currently under analysis will aid in the hazard recognition process. For example, by reviewing the lessons from an earlier spacecraft system, it might be discovered that the use of a particular seal design on a specific part of the spacecraft resulted in several mishaps. Seal design on this part of the spacecraft should then be recognized as a potential hazardous area requiring special focus in current new spacecraft design. Historical data is a valuable asset that can aid in the identification of hazards. Some common sources of useful information include the following:

- Mishap/accident/ reports
- Incident reports
- Lessons learned databases
- Customer complaints
- Maintenance and repair records

Much can be learned from historical failure data, past mistakes and lessons learned. We are doomed to repeat past mistakes if we do not keep a record of them and continuously study and apply them in HA. However, there must be an established documented protocol and procedure to track hazards in order to use said information in the evaluations of incidents.

Hazard Tracking Tools

Hazard tracking is the process of systematically recording all identified hazards and the data associated with these hazards as they progress through the lifecycle of a hazard: identification, assessment, mitigation, verification, acceptance and closure. It is typically achieved through the use of a formal electronic hazard tracking system (HTS). An effective HTS actually establishes a process that facilitates record keeping, generating reports and maintaining the necessary evidence of a safety program. A well designed HTS is an effective tool for the HA process and the SSP.

9.3 HA Tools vs. HA Tasks

Table 9.1 lists the various HA tools according to list of HA tasks established in chapter 7.

Table 9.1 – HA Tasks vs. Tools

	Steps	Tasks	Tools
1	Plan HA	• Evaluate HA requirements • Select HA technique • Establish ground rules • Establish risk criteria	- HA plan - HA ground rules - Risk criteria - Safety standards
2	Understand System Design	• Obtain design data • Understand design	- Drawings - Simplified diagrams - FFDs / RBDs - Design documents
3	Acquire HA Tools	• Identify tools • Acquire tools	- Hierarchy table - SMM - TLMs
4	Identify Hazards	• Apply HA technique • Follow ground rules • Recognize hazards • Test credibility	- HA technique - System hierarchy table - Hazard checklists - Mishap reports - FMEAs - SE diagrams
5	Validate Hazards	• Peer reviews • SSWGs	- Meetings - HTS
6	Assess Risk	• Obtain failure rate data • Obtain severity data • Calculate risk	- Hazard risk matrix - Failure data - FMEAs
7	Mitigate Risk	• Develop hazard defenses • Establish safety design requirements	- Safety requirements
8	Verify Mitigation	• Test result reports • Requirements verification	- Testing - SSWGs
9	Accept Risk	• Package hazards • Obtain approvals	- Risk matrix - SSWGs
10	Track Hazards	• Record hazards • Track hazards/risks • Close hazards	- HTS - SSWGs

9.4 Hazard Tracking System (HTS)

Hazard tracking is the process of systematically recording all identified hazards and the data associated with these hazards as they progress through the lifecycle of a hazard: identification, assessment, mitigation, verification, acceptance and closure. It is typically achieved through the use of a formal HTS. Hazard tracking is a basic required element of an effective system safety program (SSP). Hazard tracking encompasses the hazard analysis process, the mishap risk management process and the hazard mitigation process, and it ensures that no hazards are lost or overlooked.

A HTS is a tool for formally tracking all identified hazards within a system. A HTS ensures that identified hazards are properly mitigated and closed, and that all related actions are recorded; it establishes a process that facilitates hazard control through the steps of mitigation design, mitigation verification, risk acceptance and closure. Closed-loop hazard tracking is a basic required element of an effective system safety program (SSP). An HTS enforces a formal and systematic process that ensures identified hazards are resolved and also provides a historical record. It encompasses the hazard analysis process, the mishap risk management process and the hazard mitigation process. The HTS is sometimes referred to as a hazard log, hazard database or hazard tracking database (HTDB).

An HTS does not imply that a hazard is just passively stored in a database and then forgotten. Hazard tracking is a dynamic process in which the SSP takes positive steps to eliminate or mitigate the hazard and record all actions. Hazards are tracked from inception (identification) to closure, with focus on reporting and acceptance of the final residual hazard-mishap risk. Hazard tracking should be a "closed-loop" process, meaning that the review and mitigation process is repeated iteratively, until final closure of the hazard is achieved.

The primary objectives of an HTS include:

- Retain a record of all data and tasks associated with identifying and resolving hazards
- Provide a mechanism and discipline for tracking hazards from inception through closure
- Ensure that all identified hazards are adequately mitigated (i.e., none are lost)
- Meet the SSP requirement for hazard tracking and closure

The HTS provides a process for risk mitigation, as well as maintaining a complete record and history of every identified hazard. The hazard status is maintained as "open" until it has been verified that the appropriate safety requirements for eliminating or controlling the hazard have been implemented and proven successful through testing. Following successful verification and validation of the system safety requirements for a hazard, and acceptance of the hazard risk level by the appropriate authority, the hazard's status can be changed to "closed". Note that this requires documented evidence. Note that hazard closure does not imply zero risk; it means that the risk has been duly mitigated and accepted.

In a HTS, the database can be a manual or computerized system; however, it is highly recommended that an automated electronic database be utilized, particularly for medium and large system development programs. In addition, there are commercial electronic software HTS packages available that are already set up specifically for hazard tracking. An automated electronic database provides many advantages, such as:

- Hazard data entry is easy and efficient
- Data updates and changes are simple and efficient
- Capability exists to search for specific items based on different queries
- Capability exists to provide custom reports
- Capability exists to place on Network or Internet for access by many users
- Programming is not required if purchased
- Format can be utilized as a company standard for several different projects

Some basic considerations to address when designing, procuring and operating an HTS include the following:

- Rules for opening, monitoring and closing a hazard
- A hazard numbering scheme
- A standard format that remains consistent
- Rules for who can enter data into the HTS
- Rules for who can modify or remove data from the HTS
- The capability to generate various reports

9.5 Summary

Obtaining and utilizing HA tools that are available is an important aspect in HA. The analyst should thoroughly understand these tools, know how to obtain them and know how to apply them.

CHAPTER 10

HAZARD ANALYSIS TECHNIQUES

10.1 Introduction

In the system safety discipline there are many HA techniques that exist, along with techniques postulated as HA techniques. This chapter focuses on the HA techniques used by system safety. It provides an introduction to the concepts associated with HA techniques, and it provides a brief discussion of the more important HA techniques.

10.2 List of Major HA Techniques

Table 10.1 contains a list of the major HA techniques utilized in the system safety discipline. This list is provided at the front of this chapter in order to identify the techniques and their acronyms. The various technique parameters in the table will be explained in the following sections. These parameters include the categories of:

- Primary technique or secondary technique
- If the methodology is also recognized as a HA type
- Qualitative, quantitative or both

The "Prob" category indicates if the probability calculation for the technique is qualitative (QL), quantitative (QT) or both (B). Many of the techniques are identified as qualitative, which means they are primarily qualitative in nature. However, they could easily become quantitative if more research and probabilistic math is applied to each hazard. All of the methods are a HA techniques; the "Type" category means they are considered as both a type and a technique.

Table 10.1 – Hazard Analysis Type vs. Technique

Technique	Acronym	Primary	Secondary	Type	Prob
Preliminary Hazard List	PHL	X		X	QL
Preliminary Hazard Analysis	PHA	X		X	QL
Subsystem Hazard Analysis	SSHA	X		X	QL
System Hazard Analysis	SHA	X		X	QL
Operations and Support Hazard Analysis	O&SHA	X		X	QL
Health Hazard Assessment	HHA	X		X	QL
Functional Hazard Analysis	FHA	X			QL
Hazard and Operability Analysis	HAZOP	X			QL
Threat Hazard Assessment	THA	X			QL
Fault Tree Analysis	FTA		X		B
Event Tree Analysis	ETA		X		B
Failure Mode and Effects Analysis	FMEA		X		B
Bent Pin Analysis	BPA		X		QL
Sneak Circuit Analysis	SCA		X		QL
Safety Requirements/Criteria Analysis	SRCA		X	X	QL
Barrier Analysis	BA		X		QL
Interlock Analysis			X		B
Code Safety Analysis			X		QL
Particular Risk Analysis			X		QL
Markov Analysis	MA		X		QT
Test Hazard Analysis	THA		X		QL
Common Cause Failure Analysis	CCFA		X		B
Probabilistic Risk Assessment	PRA		X		QT
Safety Assessment Report	SAR	---	---		QL

Prob = Probability calculation is QL (qualitative), QT (quantitative) or B (both)

10.3 HA Type vs. Technique

Some confusion has resulted from the development of the concept of HA *types* and HA t*echniques*. HA *type* defines an analysis category or class of analysis, whereas the HA *technique* defines a unique analysis methodology. In general, HA type defines the "what and when" to analyze for safety, while the HA technique defines the specific "how to" perform the analysis. Refer to chapter 10 for more detail.

The overarching distinctions between type and technique are summarized in Table 10.2.

Table 10.2 – Hazard Analysis Type vs. Technique

TYPE	TECHNIQUE
• Establishes where, when and what to analyze • Establishes a specific analysis task at specific time in program lifecycle • Establishes what is desired from the analysis • Provides a specific design focus	• Establishes how to perform the analysis • Establishes a specific and unique analysis methodology • Provides the information to satisfy the intent of the analysis Type

System safety is built upon seven basic HA *types* originally established in MIL-STD-882, which include: PHL, PHA, SSHA, O&SHA, HHA and SRCA. Each of these analysis types define a point in time when the analysis should begin, the level of detail of the analysis, the type of information available and the analysis output. The goals of each analysis type can be achieved by various analyses techniques. The analyst needs to carefully select the appropriate techniques to achieve the goals of each of the analysis types.

An important principle about HA is that one particular HA type does not necessarily identify all the hazards within a system; identification of hazards may take more than one HA type, hence the seven types. A corollary to this principle is that one particular HA type does not necessarily identify all of the hazard causal factors; more than one HA type may be required. After performing all seven of the HA types, all hazards and causal factors should have been identified, assuming an adequate analysis program was conducted. Additional hazards that were overlooked may be discovered by other means, such as the test program.

HA *technique* refers to a specific and unique analysis methodology that provides specific results and is performed according to an established set of rules or guidelines. The overarching distinctions of a HA technique includes the following:

- Establishes "how to" perform the analysis methodology
- Establishes the analysis rules, guidelines and graphics
- Establishes the level of detail of the information required for the analysis
- Establishes the technical expertise required
- Provides the information needed to satisfy the intent of a particular analysis Type

HA techniques can have many different inherent attributes, which makes their utility different. The appropriate technique to use can often be determined from the inherent attributes of the methodology itself. The following is a list of the most significant attributes for a HA methodology or technique:

- Qualitative or quantitative
- Level of analysis detail
- Data required for the analysis
- Time required to perform the analysis
- Inductive or deductive approach
- Complexity of the analysis
- Difficulty of the analysis
- Technical expertise required to perform the analysis
- Tools required to support the analysis
- Cost of the analysis
- Subjective capability of the method

The primary confusion factor is that the seven HA types also have corresponding HA techniques by the same name. This is unfortunate, but the main objective is to focus on techniques rather than types. Do not be overly concerned about types vs. techniques. Select the HA technique, or techniques, for your program based on the program objectives, needs and customer requirements. Use the HA techniques that will best make your system or product safe.

10.4 Primary HA vs. Secondary HA

There is some confusion regarding which HA techniques are truly a HA and which are not. Within the system safety discipline there are over 100 different HA techniques that have been proposed, some of which are unique, some of which are variants of others, some of which are extremely useful and some of which are not useful at all. Some of the techniques are not true HAs, and many are merely variations of other HA techniques. There are only about 15 to 20 HA techniques that are commonly used by system safety experts.

Essentially, there are *primary* and *secondary* HA techniques. The primary HA techniques are full-fledged, or complete, formal methodologies that are designed for identifying all, or most, system hazards. The secondary HA techniques are limited in their hazard identification ability; typically they are not designed to identify all hazards. The secondary techniques essentially provide support for the primary techniques; many help identify the root causal factors of already identified hazards.

It should be noted that many analysis techniques are incorrectly used for HA, such as Failure Mode and Effects Analysis (FMEA). FMEA results can be used as resource information for an HA, but the FMEA does not suffice for an HA because it does not thoroughly cover system hazard-mishap scenarios and it does not cover the combined effect of multiple simultaneous failures that often cause hazards. An FMEA is not a HA or a safety analysis and should not be used in place of either, but it can be used to supplement them as a secondary supporting technique; it is an excellent source for failure modes and failure rates.

10.5 HA Technique Descriptions

Descriptions of the most used HA techniques are provided in the following paragraphs. For a more complete and thorough description of these techniques, refer to reference 1, which contains complete descriptions and examples for each.

10.5.1 Primary HA Techniques

Preliminary Hazard List (PHL)

The PHL is a primary type HA technique; performed very early in the system design to establish basic hazards, hazard categories, safety-critical functions (SCFs), TLMs and major safety concerns. The results are used as input when developing a system safety program plan, and as input to more detailed HAs. The PHL is a very high level analysis that covers the entire system; the amount of design information available is usually limited at the time of this analysis.

Preliminary Hazard Analysis (PHA)

The PHA is a primary type HA technique; performed during preliminary design to identify hazards and continue the risk assessment and hazard control processes. The PHA is an intermediate level analysis that covers the entire system. The intent of the PHA is to affect the design for safety as early as possible in the development program; it typically deals with incomplete design information (level of detail). It generally begins with hazards from the PHL hazards and then more thoroughly evaluates the entire system to identify previously unrecognized hazards early in the system development. The PHL initiates the derived design safety requirements necessary to mitigate hazards. The PHA also identifies SCFs and TLMs that that provide a safety focus for the design process. The PHA is applicable to the analysis of all types of systems, facilities, operations, and functions; the PHA can be performed on a unit, subsystem, system or an integrated set of systems. The PHA is probably the most commonly performed HA technique.

Subsystem Hazard Analysis (SSHA)
The SSHA is a primary type HA technique; performed during detailed design to identify hazards associated with the design of individual subsystems. The SSHA is an in-depth and detailed analysis of hazards previously identified by the PHA; it also identifies new hazards. It requires detailed design information and a good understanding of the system design and operation.

System Hazard Analysis (SHA)
The SHA is a primary type HA technique; performed during preliminary and detailed design to identify system level hazards. The SHA is a detailed level analysis that covers the entire system design. It covers subsystem interfaces, common-cause failures that cross subsystem boundaries and particular risks, such as fire, EMI, flooding, etc. The overall purpose of the SHA is to ensure safety at the integrated system level.

Operations and Support Hazard Analysis (O&SHA)
The O&SHA is a primary type HA technique; performed during preliminary and detailed design for identifying hazards in system operational tasks, along with the hazard causal factors, effects, risk and mitigating methods. The O&SHA is an analysis technique for specifically assessing the safety of operations by integrally evaluating operational procedures, the system design and the Human System Integration (HSI) interface. The scope of the O&SHA includes normal operation, test, installation, maintenance, repair, training, storage, handling, transportation, emergency and rescue operations. It also covers written procedures for these tasks. Consideration is given to system design, operational design, hardware failure modes, human error, and task design. Human factors and HSI design considerations are a large factor in system operation, and therefore also in the O&SHA.

Health Hazard Assessment (HHA)
The HHA is a primary type HA technique; performed during preliminary and detailed design to identify human health related hazards that could result from the system design or system operations. O&SHA is a detailed

level analysis for evaluating the human health aspects of a system's design. These aspects include considerations for ergonomics, noise, vibration, temperature, chemicals, hazardous materials, etc. The intent is to identify human health hazards during design and eliminate them through design features. If health hazards cannot be eliminated, then protective measures must be used to reduce the associated risk to an acceptable level. Health hazards must be considered during manufacture, operation, test, maintenance and disposal.

Functional Hazard Analysis (FHA)
The FHA is a primary type HA technique; performed during preliminary and detailed design to identify system level hazards. The SHA is a detailed level analysis that covers the entire system design. The purpose of FHA is to identify system hazards by the analysis of functions. Functions are the means by which a system operates to accomplish its mission or goals. System hazards are identified by evaluating the safety impact of a function failing to operate, operating incorrectly, or operating at the wrong time. When a function's failure can be determined hazardous, the casual factors of the malfunction should be investigated in greater detail. FHA is applicable to the analysis of all types of systems, equipment and software. FHA can be implemented on a single subsystem, a complete functional system or an integrated set of systems.

Hazard and Operability Analysis (HAZOP)
The HAZOP is a primary type HA technique; performed during preliminary and detailed design for identifying hazards in the system design. It was developed by the chemical process industry and applies a very formal and rigid methodology. HAZOP analysis utilizes use key guidewords and system diagrams (design representations) to identify system hazards. Adjectives (guide words) such as *more, no, less,* etc. are combined with process/system conditions such as *speed, flow, pressure,* etc. in the hazard identification process. HAZOP analysis looks for hazards resulting from identified potential deviations in design operational intent. A HAZOP analysis is performed by a team of multidisciplinary experts in a brainstorming session under the leadership of a HAZOP team leader.

10.5.2 Secondary HA Techniques

Fault Tree Analysis (FTA)
FTA is a secondary type HA technique; performed during preliminary and detailed design. FTA utilizes a visual logic tree structure; its main purpose is to identify and graphically model all of the causal factors that can lead to the occurrence of an undesired event. It is both a qualitative and quantitative analysis; it identifies causal factors and their probability of occurrence. It can also provide a top event probability of occurrence, which is used in a probabilistic risk assessment (PRA).

Event Tree Analysis (ETA)
ETA is a secondary type HA technique; performed during preliminary and detailed design. ETA is an analysis technique for identifying and evaluating the sequence of events in a potential accident scenario following the occurrence of a postulated initiating event. ETA utilizes a visual logic tree structure. The objective of ETA is to determine whether the initiating event will develop into a serious mishap, or if the event is sufficiently controlled by the safety systems and procedures implemented in the system design. An ETA can result in many different possible outcomes from a single initiating event, and it provides the capability to obtain a probability for each outcome. ETA combines a decision tree for the evaluation of multiple outcomes, with a FTA to determine the cause and probability of certain failure events in the ETA.

Failure Mode and Effects Analysis (FMEA)
FMEA is a secondary type HA technique; performed during preliminary and detailed design. FMEA is a tool for evaluating the effect(s) of potential failure modes of subsystems, assemblies, components, functions and processes. It is primarily a reliability tool to identify failure modes that would adversely affect overall system reliability. FMEA has the capability to include failure rates for each failure mode, in order to achieve a quantitative probabilistic analysis. FMEA only considers single element failures and not multiple failures; it is not geared to identify hazard scenarios. FMEA is not a formal HA technique and does not suffice for one. It can be used as a resource for a primary HA technique for the identification of critical component failure modes and failure rates.

Bent Pin Analysis (BPA)
BPA is a secondary type HA technique; performed during preliminary and detailed design; it is used for identifying hazards caused by bent pins

within cable connectors. It is possible to improperly attach two connectors together and have one or more pins in the male connector bend sideways and make contact with other pins within the connector. If this should occur, it is possible to cause open circuits and/or short circuits to +/- voltages, which may be hazardous in certain system designs. For example, a certain cable may contain a specific wire carrying the fire command signal (voltage) for a missile. This fire command wire may be contained within a long wire that passes through many connectors. If a connector pin in the fire command wire should happen to bend and make a short circuit with another connector pin containing +28 VDC the missile fire command may be inadvertently generated. BPA is a tool for evaluating all of the potential bent pin combinations within a connector to determine if a potential safety hazard exists.

Sneak Circuit Analysis (SCA)
SCA is a secondary type HA technique; performed during preliminary and detailed design. A sneak circuit is a latent path or condition in an electrical system that inhibits a desired condition or initiates an unintended or unwanted action. This condition is not caused by component failures, but by circuit paths that have been inadvertently designed into the electrical system to occur as normal operation. Sneak circuits often exist because subsystem designers lack the overall system visibility required to electrically interface all subsystems properly. When design modifications are implemented, sneak circuits frequently occur because changes are rarely submitted to the rigorous testing that the original design undergoes. Some sneak circuits are evidenced as "glitches" or spurious operational modes and can be manifested in mature, thoroughly tested systems after long use. SCA can be applied to both hardware and software design.

Safety Requirements/Criteria Analysis (SRCA)
SRCA is a secondary type HA technique; performed during preliminary and detailed design. The SRCA is essentially a traceability analysis to ensure that there are no holes or gaps (i.e., no hazard has been left unmitigated) in the safety requirements and that all identified hazards have adequate and proven design mitigation coverage. The SRCA applies to hardware, software, firmware design requirements. The SRCA process consists of comparing the SSRs to design requirements and identified hazards. In this way any missing safety requirements will be identified. In addition, SSR's are traced into the test requirements to ensure that all SSRs are tested. A matrix worksheet is used to correlate safety requirements with design requirements, test requirements and identified hazards. If a hazard

does not have a corresponding safety requirement, then there is an obvious gap in the safety requirements. If a safety requirement is not included in the design requirements, then there is a gap in the design requirements. If a safety requirement is missing from the test requirements, then that requirement cannot be verified and validated. If an SSR cannot be shown to have passed testing, then the associated hazard cannot be closed.

Barrier Analysis
Barrier Analysis is a secondary type HA technique for identifying and mitigating hazards specifically associated with hazardous energy sources. BA provides a tool to evaluate the unwanted flow of hazardous energy to targets, such as personnel or equipment, and the evaluation of barriers preventing or reducing the hazardous energy flow. Many system designs cannot eliminate energy sources from the system since they are a necessary part of the system. The purpose of BA is to evaluate these energy sources and determine if potential hazards in the design have been adequately mitigated through the use of energy barriers. It is a secondary HA technique.

Interlock Analysis
An Interlock Analysis is a secondary type HA technique. An interlock is a design safety feature whereby the operation of one control or mechanism allows, or prevents, the operation of another function based upon a set of pre-determined safety criteria. It should be noted that Interlocks are absolutely not necessary for the operational functionality of a system. An interlock analysis evaluates the design and failure modes associated with a particular interlock design. The interlock analysis is also referred to as an Interlock Hazard Analysis.

Threat Hazard Assessment (THA)
THA is a secondary type HA technique. The THA is an evaluation of a munition and its life cycle profile to determine the threats and hazards to which the munition may be exposed. The assessment includes threats posed by friendly munitions, enemy munitions, accidents, handling, transportation, storage, etc. The THA identifies threats and hazards, both qualitatively and quantitatively, along with their causes and effects.

Test Hazard Analysis
Test Hazard Analysis is a secondary type HA technique. It is an analysis that identifies hazards associated with testing to ensure that all planned testing is safe. The objective is to evaluate the test procedures to ensure

that the procedures as written are not hazardous, and that the procedures provide warning and cautions for all hazards involved in the testing. Test HA evaluates each test, the system design and potential human error. It also includes coverage of any special test equipment that is used during testing. Test Hazard Analysis can apply to all levels of testing, such as component, assembly, unit, subsystem and system. Test hazard analysis is typically a secondary HA; however, it could be primary for a test program, but not system design.

Common Cause Failure Analysis (CCFA)
CCFA is a secondary type HA technique. A Common Cause Failure (CCF) is a single failure or fault that causes multiple system elements to fail simultaneously. For example, flooding can cause multiple simultaneous failures of equipment in the flood zone. CCFA is an analysis that identifies potential CCFs that can produce system hazards. A major concern in CCFA is the simultaneous failure of safety-critical redundant items.

Probabilistic Risk Assessment (PRA)
A PRA is a secondary type HA technique. A PRA is a quantitative evaluation that is performed to determine the probabilistic risk associated with an event, typically involving a complex system. The PRA provides the probability of an event occurring and the overall consequential severity of the event. Performing a PRA requires a systematic and comprehensive methodology, such as FTA. A PRA answers the questions of a) what are the possible undesired outcomes, b) what are the root causal factors involved and c) what is the risk probability presented.

10.6 Related HA Techniques

In the system safety discipline there are some reports that are well known and often mistakenly thought to be HAs. The following fall into this category.

Safety Assessment Report (SAR)
A SAR is not a HA, but a summary report. A SAR is a comprehensive report that provides a risk assessment of the system and evidence of the system safety program effectiveness. The SAR is a snapshot of the potential mishap risk a system design presents at a particular point in time in the program. The SAR is a living or evolving document, updated at each program milestone to reflect the current mishap risk status. The SAR is

built upon HAs; it matures in levels of detail as the system design progresses and matures.

10.7 Summary

There are many different HA techniques available for the safety analyst. Quite often several techniques are needed to ensure complete system and operational coverage. It is important to select the correct HA techniques for the job and to fully understand the technique. It is also important to understand the difference between primary and secondary HA techniques and to ensure safety risk assessments are based only on primary techniques and not secondary techniques.

CHAPTER 11

HAZARD RECOGNITION CHECKLISTS

11.1 Introduction

Hazard checklists are a valuable resource when performing a HA. The purpose of a hazard checklist is for the analyst to use the checklist as guidewords to help recognize hazards in a particular system. Hazard checklists contain keywords that are used to trigger hazard ideas. A safety analyst should have as many hazard checklists at his/her disposal as possible. A good safety analyst should have a personal collection of hazard checklists. It does not matter if some items in different checklist overlap, the idea is to provide as much coverage as possible.

This chapter focuses on some example hazard checklists. It is recommended that not only safety analysts use these checklists, but that managers also use checklists when reviewing a HA to ensure a thorough analysis has been performed.

11.2 Example Hazard Checklists

One of the first steps in performing a HA is to acquire the appropriate hazard checklists. Hazard checklists are generic lists of items known to be hazardous or that might create potentially hazardous designs or situations. The hazard checklist should not be considered complete or all-inclusive. Hazard checklists help to trigger the analyst's recognition of hazards. Hazard checklists are typically derived from experience and past lessons learned. Typical hazard checklists include:

- a) Energy sources
- b) Hazardous functions
- c) Hazardous operations
- d) Hazardous components
- e) Hazardous materials
- f) Lessons learned from similar type systems
- g) Undesired mishaps

h) Failure mode and failure state considerations

Hazard checklists provide a common source for readily recognizing hazards. Since no single checklist is ever really adequate in itself, it becomes necessary to develop and utilize several different checklists. Utilizing several checklists may generate some repetition, but will also result in improved coverage of hazardous elements.

Remember that a checklist should never be considered a complete and final list, but merely a mechanism or catalyst for stimulating hazard recognition. To illustrate the hazard checklist concept, some example checklists are provided in Tables 11.1 through 11.6. These example checklists are not intended to represent ultimate checklist sources, but some typical example checklists used in recognizing hazards; they range from the simple checklist to more complex checklists.

Table 11.1 is a checklist of energy sources that are considered to be hazardous elements when used within a system. The hazard is generally from the various modes of energy release that are possible from hazardous energy sources. For example, electricity/voltage is a hazardous energy source. The various hazards that can result from undesired energy release include personnel electrocution, ignition source for fuels and/or materials, sneak path power for an unintended circuit, etc.

Table 11.1 – Hazard Checklist for Energy Sources

Actuating devices	Nuclear material/devices
Catapulted objects	Pressure containers
Charged electrical capacitors	Propellants
Cryogenics	Pumps, blowers, fans
Electricity/voltage	RF/EMR energy sources
Explosives	Radioactive energy sources
Falling objects	Rotating machinery
Fuels	Spring-loaded devices
Gas generators	Static electrical charges
Heating devices	Storage batteries
Initiators	Suspension systems

Table 11.2 is a checklist of general hazard sources that are considered to be hazardous elements when used within a system. Some of these sources also include energy sources.

Table 11.2 – Hazard Checklist for General Sources

Acceleration	Noise
Chemical dissociation	Non-Ionizing Radiation (EMR)
Chemical replacement	Oxidation
Contamination	Pressure
Corrosion	Shock (mechanical)
Electricity / Voltage	Stress concentrations
Explosion	Stress reversals
Fire	Structural damage or failure
Heat	Temperature (variations)
Ionizing Radiation (nuclear)	Toxicity
Leakage	Vibration
Moisture	Weather

Table 11.3 is a checklist of general space functions that are considered to be hazardous elements for space and spacecraft operations.

Table 11.3 – Hazard Checklist for Space Functions

Crew egress/ingress	Loss of power/control
Crew recovery	Mid-course correction
Data acquisition	On-orbit performance
Explosion	Orbit correction
Extra vehicular activity	Orbit positioning
Fairing separation	Orbital injection
Fire / Toxicity	Parachute deployment and descent
Ground control communication	Payload mating
Ground control of crew	Re-entry
Ground data communication to crew	Rendezvous and docking
Ground to stage power transfer	Retro-thrust
In-flight emergencies	Solar panel deployment
In-flight tests by crew	Stage firing and separation
Launch escape	Star acquisition (navigation)
Life support	Vehicle inerting and decontamination
Loss of communications	Vehicle safing and recovery

Table 11.4 is a checklist of general functions and operations that are considered to be hazardous elements when used within a system.

Table 11.4 – Hazard Checklist for General Operations

Cleaning
Extreme temperature operations
Extreme weight operations
High energy pressurization/hydrostatic-pneumostatic testing
Hoisting, handling, and assembly operations
Manned vehicle tests
Nuclear component handling/checkout
Ordnance installation/checkout/test
Proof test of major components/subsystems/systems
Propellant loading/transfer/handling
Static firing
Tank entry/confined space entry
Test chamber operations
Transport and handling of end item
Welding

Table 11.5 is a checklist of possible failure modes or failure states that are considered hazardous, depending on the critical nature of the operation or function involved. This checklist is a set of key questions to ask regarding the state of the component, subsystem or system functions. These are potential ways the subsystem could fail and thereby result in creating a hazard. For example, when evaluating each subsystem, answering the question "does *fail to operate* cause a hazard" may lead to the recognition of a hazard.

Table 11.5 – Hazard Checklist for Failure States

Fails to operate
Operates incorrectly/erroneously
Operates inadvertently
Operates at incorrect time (early, late)
Unable to stop operation
Receives erroneous data
Sends erroneous data

Table 11.6 is a generic hazard checklist of generic hazard sources and the hazardous consideration for each.

Table 11.6 – Generic Hazard Checklist

Hazard Source	Safety Considerations	Safety Thoughts
Actuating Device	Failure to actuate Incomplete actuation Inadvertent actuation	Failure to control device Inadvertent device operation
Chemicals	Fluid intrusion Fire / explosion Caustic damage	CCF
Common Cause Failure (CCF)	Bypasses redundancy Causes system failure Common mode exposure Particular risk exposure Proximity exposure Zonal exposure	Close proximity Dust / Dirt Fire Flooding Maintenance error Manufacturing error Moisture/Humidity Power outage Radiation Same zone or location Seismic disturbance Temperature Vermin / Varmints Vibration
Control System	Fails to operate Operates prematurely Operates inadvertently Operates erroneously	System failure Unable to control system Erroneous system operation SCFs are critical
Cryogenics	Fails when needed Leakage	Death / injury Equipment failure
CRT Display	Explosion Implosion High voltage	Display errors on SCFs
Electricity	Bent pins in connectors (shorts) Electric shock / burns Electrical short Electrocution Overheating Power not available	Exposed contacts / wires Ignition source Inadvertent activation System power source (CCF)
Environment	Earthquake Firestorm Flooding Ice Lightning Rain / snow / hail RF energy Tornados / hurricanes Wind storm	Intruders Other systems
Ergonomic	Fatigue Glare Lighting Noise Temperature	Sitting position Lifting Workstation design Controls positions Controls readability
Explosions	Explosive dust Explosives Fuel Propellant	Confinement Ignition source Separation Ventilation
Explosives	Blast overpressure Explosion Fire Flying fragments	Electrostatic discharge Heat / cold Inadvertent ignition Induced voltage Lightning Low humidity RF energy

		Shock / impact / vibration Sympathetic detonation
Fire / Flammability	Fuel Ignition source Oxidizer	Chemicals Explosives Flammable material Propellant
Fluids	Water intrusion Flooding Back flow / siphon effect	CCF Connection failures Storage failures
Fuels	Detection system fails Explosion Fire Suppression system fails	CCF Fire extinguisher Leakage Toxic fumes
Hazardous Material (HM)	Personnel death/injury Environmental damage	Refer to HM data sheets
Height	Trips and falls Falling equipment	Barriers Protective equipment
Human	Erroneous operation Failure to operate Human confused by design Human error Inadvertent operation Rule violation Sequential error Unable to shutdown equipment	Drugs/alcohol Illness Instruction error Rest Stress Training
Hydraulic	Pressure Fluid	Fire combustible Inadvertent release Line leak / rupture Pressure relief fails Tank rupture
Ionizing Radiation	Alpha Beta Gamma Neutron X-ray	Excessive exposure Inadvertent exposure Inadvertent release Unknown exposure
Leaks / Spills	Liquids Chemicals	Asphyxiating Corrosive Fire hazard Flood hazard (CCF) Slip hazard Toxic
Lightning	Electrocution / injury Equipment damage Fire ignition source	
Maintenance	Calibration Instruction errors Timing Tools Training	
Mechanical	Crush surfaces Ejected parts Lifting weight Pinch points Sharp edges	Structural failure
Mission Phases	Hazardous phases Inadvertent entry Inadvertent exit Phase transitions	Environment Storage Transport
Motion	Acceleration De-acceleration Rotating Linear Repetitive motion tasks	Fails when needed Occurs inadvertently Items caught in movement Flying objects Injury
Noise	Excessive noise level Noise exceeds standards	Hearing damage Hearing protection

Non-ionizing Radiation	Laser Microwave Infrared Ultraviolet RF energy	Excessive exposure Inadvertent exposure Unknown exposure
Physiological	Allergens Asphyxiants Carcinogens Fatigue Irritants Lifted weights Lighting Mutagens Noise Radiation Repetitive motion Stress Temperature Toxins Vibration (Raynaud's Syndrome)	Confusion Death Injury Sickness
Pneumatic	Air Pressure	Fire oxidizer Inadvertent release Line leak / rupture Pressure relief fails Tank rupture
RF Energy	Ignition source Induces inadvertent signals Interference / jamming	
Rotating Device	Disintegration Items caught in device	Failure to control device Inadvertent device operation
Safety Critical Function (SCF)	Fails to operate Operates erroneously Operates inadvertently Operates prematurely	Erroneous system operation System failure Unable to control system
Safety System	Fails to operate when need Coverage is inadequate	CCFs with main system
Software	Causes erroneous command Causes inadvertent command Causes incorrect display output Causes sequence errors Corrupts data Generates unintended function	Design rigor Protect safety-critical data Protect SCFs Test rigor
Temperature	Extremes Freezing Heat Normal	Burns Cooling / heating failure Elevated flammability Equipment failure Equipment failure rate reduced Embrittlement Hot surfaces Humidity / moisture
Touch Screen Display	Display data confuses operator Erroneous display of SCFs Erroneous touch command Inadvertent touch command	Screen breaks Switch to alternate screen
Vibration	Excessive vibration Resonant vibration	Structural damage CCF
Weight	Item falls Item rollover Personnel lifting injury	Lifting devices Warnings Weight limits

11.3 Summary

Hazard checklists are an effective tool in HA. Some basic guidelines for the use of hazard checklists include the following:

- No single checklist is sufficient; a good HA requires the use of many different checklists.
- No hazard checklist should be considered complete.
- Never rely on a checklist alone as the only means of identifying system hazards.
- Hazard checklists are essentially only keywords that are used to trigger hazard ideas.
- Expand and tailor your checklists as experience is gained.
- Collect and use as many checklists as possible.

CHAPTER 12

COMMON HAZARD ANALYSIS MISTAKES

12.1 Introduction

When first learning how to perform a HA it is commonplace to commit some typical errors. This chapter contains a list of errors often made during the conduct of a HA. It is important to fully understand each of these mistakes in order to prevent repeating them.

12.2 Common Mistakes

The following is a list of mistakes and errors commonly committed during HA:

1) Failure to list all postulated hazard concerns or credible hazards. It is important to list all possible suspected or credible hazards and not leave any potential concerns out of the analysis.

2) Failure to document postulated hazards that were identified, but found to be not credible. The HA is a historical document encompassing all hazard identification areas that were considered. When questions arise at a later date it can be shown that certain aspects were indeed considered.

3) Failure to utilize a structured HA approach or methodology. Always use a worksheet, and include all equipment, energy sources, functions, environments, etc.

4) Failure to collect and utilize common hazard source checklists. These checklists provide an invaluable aid in hazard identification. They are typically derived from lessons learned.

5) Failure to research similar systems or equipment for mishaps and lessons learned that can be applied to the current HA effort.

6) Failure to establish a correct and complete list of hardware, system functions, mission phases and expected environments.

7) Assuming readers of the HA will understand the hazard description from a brief and abbreviated hazard description filled with project unique terms and acronyms.

8) Inadequately describing the identified hazard. For example, insufficient detail, too much detail, incorrect hazard effect, wrong equipment indenture level, or not identifying all three elements of a hazard.

9) Inadequately describing the causal factors. For example, the identified causal factor does not support the hazard, the causal factor is not detailed enough, or not all of the causal factors are identified.

10) Inadequately describing the Hazard Risk Index (HRI). For example, the HRI is not stated or is incomplete, the hazard severity level does not support actual hazardous effects, the Final HRI is a higher risk than the initial HRI, the final severity level in the risk is less than the initial severity level (sometimes possible, but not usually), the hazard probability is not supported by the causal factors.

11) Providing recommended hazard mitigation methods that do not address the actual hazard causal factor(s) in the identified hazard.

12) Incorrectly closing a hazard by not following the procedures established by the program.

13) Prematurely closing a hazard without complete causal factor analysis and verification that mitigation methods have been implemented and successfully tested.

14) Failure to establish and utilize a list of TLMs and SCFs for the system under investigation.

15) Failing to thoroughly investigate the hazard causal factors to ensure that everything is considered and all causal factors have been identified.

16) Applying a Mishap Risk Index (MRI) risk severity level that does not appropriately support the identified hazardous effects and outcomes.

17) Failure to consider common cause events and dependent events. Not conducting a thorough investigation of CCF factors, events or groups. Not evaluating all redundant subsystems for CCF vulnerability. Not using a FTA for visualization of CCF events.

18) Failure to identify supporting data and analyses in the HA worksheets.

19) Failure to establish a HA plan; the plan should include HA ground rules, methodologies and guidelines critical to the analysis.

20) Failure to establish analysis rules or guidelines that state at what level in the system hierarchy hazard risk levels should be developed.

21) Failing to provide the exposure time used in probability calculations. Unscrupulous analysts may use a smaller than valid exposure time in order to reduce the probability to an acceptable level, and fail to divulge this fact.

22) Failure to establish how hazards should be described. The hazard description should include the entire hazard scenario and include the HS-IM-TTO elements.

23) Failure to have the HA reviewed by subject matter experts (SMEs) for accuracy in their field of expertise.

24) Failure to establish a representative hazard architecture and hazard hierarchy in the form of a SMM.

12.3 Summary

HA is not a trivial process; it is fraught with complexities and misunderstandings. Mistakes can be avoided by understanding lessons learned from this chapter.

CHAPTER 13

HAZARD CAUSAL FACTORS

13.1 Introduction

There is a difference between *why* hazards exist and *how* they exist. The basic reasons *why* hazards exist are: (1) they are unavoidable because hazardous elements must be used in the system, and/or (2) they are the result of inadequate design safety consideration. The reasons *how* hazards exist refers to the hazard causal factor (HCF) involved, or the specific set of HCFs involved. These include factors such as specific and discrete component failures, specific human errors, specific adverse environmental conditions, etc. that adversely affect the system. Understanding HCFs from both a why and how perspective is important in effective HA.

Hazards and HCFs are classified in many different ways. Different classifications have been developed by people attempting to better understand hazards. Some of these classifications are useful, some are useless and some are incorrect; care must be taken.

When looking at hazards from a very high level perspective, they are basically caused by design flaws, component failures and human error. However, in order to identify hazards and mitigate them, HCFs must be broken down and understood at a much lower level. This chapter focuses on HCFs and the assorted classifications and concepts associated with them.

13.2 Postulated Hazard vs. Actual Hazard

When identifying hazards during a HA, there are typically two stages involved. In the first stage the hazard scenario is postulated, or put forth for consideration. It is a hazard concept that seems to make sense and seems to be relevant to the system under analysis. In the second stage, the postulated hazard is analyzed in more detail and determined to be either a valid hazard or a non-valid hazard. Sometimes after due consideration it is realized that a postulated hazard is not credible; it does not really make sense or it does not truly exist. Sometimes in a HA it is important to document or record all postulated hazards that were determined to be non-relevant. This provides a record that everything was considered in the HA, including items that turned out not to be hazards.

The bottom line is: a hazard either exists or it doesn't, there is no in between or maybe. But, it is important to record everything that was considered, in case someone at a later point in times asks "did you consider such and such?"; it can be shown that it was considered and the HA was complete and thorough.

13.3 Plausible Hazard vs. Real Hazard

Although a hazard is a real entity that exists within the design of a system, it is somewhat invisible; that is, it is camouflaged or concealed by the overall system complexity. Some hazards are much easier to recognize or visualize than others. In this regard, there are *real* hazards and *plausible* hazards.

In order for a hazard to exist there must be a Hazard Source (HS), an Initiating Mechanism (IM) and an undesired Target/Threat Outcome (TTO) for the hazard. When identifying hazards the HS and TTO are the most easily identifiable elements, however, the IM element is a little more elusive. When dealing with hardware related hazards the specific IM can always be identified. In software related hazards the specific IM is usually not identifiable. In some respect, these two types of hazards could also be thought of as "hard hazards" vs. "soft hazards" for this reason.

A real hazard is a hazard where all three of the required elements can be identified in sufficient detail to obtain data for a risk assessment. A real hazard is an actual concrete hazard where all of the causal factors can be concretely established. For example, the specific IMs for a toaster fire hazard would be failure of the browning turn-off timer and failure of the over-temp sensor. The specific IMs for a missile inadvertent launch hazard are failure of both the Fire-1 and Fire-2 buttons that are necessary for launch.

A plausible hazard is a hazard where all of the hazard causal factors cannot be firmly established, yet the hazard can be defined at a high level of abstraction. For example, consider the hazard: Inadvertent missile launch due to a software error. The hazard is credible, but the specific software cause may not be identifiable.

Plausible hazards can cause confusion in HA, and obtaining a risk assessment for them can be difficult, if not impossible. Thus, plausible hazards and real hazards will sometimes co-exist in a HA.

13.4 Software Hazard vs. Hardware Hazard

There is a difference between *hardware* hazards and *software* hazards. As the name implies, hardware hazards involve system hardware items, which tend to have tangible physical failure modes that can be identified. For example, a resistor has two primary failure modes: fail open and fail shorted. These are concrete physical failures that have failure rates associated with them, which can be used to calculate probability of failure in a risk calculation. Human error hazards and environmental hazards tend to fall into the hardware hazard category because discrete failure modes and failure rates can be identified.

In contrast, not all of the causal factors in a software hazard are concrete or uniquely identifiable since software is abstract in nature. In a software related hazard, the software has to erroneously impact hardware functions in order to trigger the hazard. Software hazards present a risk enigma for system safety; the probability of occurrence cannot be calculated, thus a risk calculation cannot be made. The typical system safety process for risk assessment of hardware hazards cannot be directly applied to software hazards. For example, an analyst can correctly postulate that the software in a missile system could inadvertently generate a missile fire command, yet the exact causal factors in the software code can never be comprehensively identified by HA for this hazard. However, the potential outcome is significant enough that the hazard cannot be discarded as non-credible, even though the specific causal factors are not known and the probability cannot be established.

Risk is the probability of a hazard occurring multiplied by the amount of loss caused by the hazard mishap. The probability portion of hazard risk for a software caused hazard cannot be established because a failure rate or probability cannot be determined for the software causal factors involved. The probability cannot be calculated since the discrete causal factors cannot be identified, and because failure rates for discrete software errors do not exist. This means that the improvement in residual risk from design mitigation cannot be estimated. In a hardware caused hazard the HCFs can be concretely identified and initial risk and final residual risk can be computed.

13.5 Hazard Categories

Analysts attempt to identify hazards by different hazard categories; the idea being that if all hazards can be classified by a category, then they will be much easier to recognize during HA. The problem is that hazard categories are generalizations, and should not necessarily be quoted as the only hazard types possible. The different categories provide different perspectives, which are useful, but not ironclad.

This section contains several different hazard classifications or classification methods. These classifications may help the safety analyst. Some hazards may fall into more than one category. But, these categories may raise questions, for example, are hazards with a system focus significantly different than those with a subsystem focus?

13.5.1 Hazard Types by General Circumstances

One classification for hazard types falls into the category of general system circumstances. Some in this category overlap with other categories. The following are some of these types of hazards:

1) Inherent Hazard – These are hazards resulting from the inherently hazardous nature of the components, equipment or processes in the system, such as hazardous materials, energy sources or safety critical functions. This probably accounts for the majority of system hazards.

2) Software Hazard – These are hazards resulting from software and its interface and control of system functions. The software associated with these system-level functions is safety-related, and errors in these functions may be a hazard causal factor. For example, missile launch is a safety critical function that has serious safety implications, such as inadvertent launch.

3) Timing Hazard – These are hazards resulting from errors or faults in areas where timing is safety-critical. These hazards can result from hardware, software and/or human performance. This is often an overlooked area because timing is often taken for granted to be safe, until an accident or mishap occurs.

4) Hardware Induced Hazard – These are hazards resulting from software errors caused by a computer hardware failure that causes a bit error, resulting in an erroneous instruction. For example, an intended word with bit pattern "1101" that means "add to register A" may be changed by an induced bit error to "1001" that means "subtract from register A." This type of hazard has safety implications in electronic data storage and transfer methods. Hardware failures can modify software and/or induce totally unpredictable results in the software and in system timing. For example, a hardware bit perturbation can result in an incorrect

instruction or data, or a jump to an incorrect memory location. Hard hardware failures can be traced and corrected, but soft failures (e.g., alpha particle radiation) occurs at random and are usually not repeatable when debugging

5) Latent Hazard – These are hazards resulting from a hidden condition in the hardware/software design which is not hazardous until a particular unanticipated, unplanned or untested set of circumstances occur. It could also be the result of a built-in unintended function or the result of a sneak path in the software. This is somewhat of a misleading category, because in reality all hazards are essentially latent.

6) Systemic Hazard – These are hazards caused by the systemic system design. An error in the hardware-software design, integration or implementation that results in a system level hazard. These types of hazards are generally the result of hardware-software interface flaws.

7) Code Error Hazard – These are hazards caused by a software coding error.

8) Explosives Hazard – These are hazards caused by the use of explosives or explosive materials in the system.

9) Common Cause Hazard – These are hazards caused by faults and failures that cause redundant system elements to fail simultaneously. For example, flooding can cause multiple simultaneous failures.

10) Sneak Circuit Hazard – These are hazards caused by sneak, or unintended, electrical paths through system circuits that perform unwanted functions.

11) Human System Integration (HSI) hazard – These are hazards caused by the user interface design. For example, the system operator can be forced into committing errors when he is confused by the poor layout of a multitude of switches, gauges and displays.

12) Organizational hazard – These are hazards caused by faults, failures or errors in an organization. Typically they deal with poor, inadequate, incorrect or lack of performance in regard to safety issues.

13.5.2 Hazard Types by Analysis Category

One classification for hazards falls into the category of basic hazard types based on their analysis derivation. The following are some of the recognized analysis based types:

1) Functional hazard – Hazards that are functional in nature; typically identified from a FHA. These hazards involve functions that must be performed by the system, and the functions are premature, fail, out of tolerance, unintended, etc. For example, inadvertent missile launch is a functional failure hazard. The functional error would have to be traced back to the hardware, software or human root cause.

2) System hazard – Hazards that are systemic in nature; typically identified from a SHA. These hazards involve synergistic system relationships. They can involve more than one subsystem and interfaces and interrelationships between subsystems.

3) Subsystem hazard – Hazards involving or contained within a single subsystem; typically identified from a SSHA. They may pertain to individual subsystems or the system. Discovery methods include: SSHA, FMEA and PHA.

4) Operational hazard – Hazards involving operating and support tasks; typically derived from an O&SHA. The effect of procedures as they are integrated with the system design.

5) Health hazard – Hazards that are human health in nature; typically derived from a HHA. Hazards involving the health of system operating or support personnel. Typically a subsystem problem, but could also be a system hazard. Examples include: operator injury from repetitive motion, operator injury from excessive noise, or operator injury from silica inhalation

6) Test hazard – Hazards involving testing. Includes test procedures, system equipment, software and test support equipment. These hazards are typically derived from a Test Hazard Analysis.

7) Software hazard – Hazards that are strictly oriented to software related HCFs. These hazards are typically derived from specially performed software safety analyses.

8) Operator Hazard – These are hazards caused by operator error, operator performance, etc. These hazards are typically derived from PHAs, SSHAs, SHAs and O&SHAs.

13.5.3 Hazard Types by Major System Element

One classification for hazards falls into the category of basic hazard types based on their primary system element grouping. The following are some of the recognized system based types:

1) Hardware – Hazards that are primarily caused by hardware failures.

2) Software – Hazards that are primarily caused by software, including design errors, compiler errors, tool errors, etc.

3) Human – Hazards that are primarily caused by various human factors or Human System Integration (HSI) in the system. This primarily involves human errors, which can result from conditions such as:
    - Operator fails to perform function
    - Operator performs function incorrectly
    - Operator performs function inadvertently
    - Operator performs wrong function
    - Operation action exacerbates the results of a system failure
    - Decision errors
    - Judgment errors
    - Memory lapses
    - Fatigue, burnout, depression, drugs, alcohol

4) Environment – Hazards that are primarily caused by environmental conditions that exceed system parameters, including both internal system environments (e.g., temperature) and external environments (e.g., weather).

5) Procedural – Hazards that are primarily caused by system procedures. This includes operating procedures, test procedures, maintenance procedures, handling and transportation procedures, etc.

13.6 Hazard Causal Factor Levels

HA and hazard mitigation are key elements in the system safety process. Before a hazard can be mitigated or controlled it is logically necessary to fully understand the hazard causal factors (HCFs) involved. One of the errors typically committed during hazard mitigation is to assume that each direct HCF must be individually identified and individually mitigated at the lowest component level. This certainly is the prime goal of safety; however, sometimes it is not possible or practical.

HCFs can be viewed and dealt with at different levels of detail. The level of detail runs the gamut from the component level to the subsystem level to the functional level, where the component level is the lowest level and the functional level is the highest level of detail. Depending upon the particular application and hazard it is sometimes more practical and effective to apply mitigation methods at the higher level than the lower level.

The basic goal would be to mitigate each individual causal factor that can be identified at the lowest level. However, there are some problems associated with this approach. Sometimes the specific HCF cannot be identified, particularly when dealing with software. Hazards can be identified with an understanding of high level HCFs but the detailed specific causal factors may not be identifiable within the actual code. Also, sometimes it is not possible to mitigate the actual specific HCF itself for various reasons and another method of mitigation surrounding the HCF is necessary.

Figure 13.1 depicts the overall HCF Model. This model correlates all of the factors involved in hazard-mishap theory. The model illustrates that hazards create the potential for mishaps, and mishaps occur based on the level of risk involved (i.e., hazards and mishaps are linked by risk). The three basic hazard components define both the hazard and the mishap. The three basic hazard components can be further broken into major hazard causal factor categories, which are 1) Hardware, 2) Software, 3) Humans 4) Interfaces, 5) Functions and 6) the Environment. Finally, the causal factor categories are refined even further into the actual specific detailed causes, such as a hardware component failure mode.

This HCF Model illustrates how hazard HCFs can be viewed at four different levels:

Level 1: The three hazard components (HS, IM, TTO).
Level 2: The main HCF categories (hardware, software, human, etc.).
Level 3: The generic causes (failure to function, etc.).
Level 4: The detailed specific causes (resistor failure, etc.).

The top level (Level 1) hazard HCF categories define the basic root cause sources for all hazards. The first step in HCF identification is to identify the category, and then identify the detailed specifics in each category, such as specific hardware component failures, operator errors, software error, and the like. The second level identifies the primary HCF categories, which include hardware, software, human, environment, functions and interfaces. The third and fourth levels break down the HCFs even further. The third level establishes causal modes, such as failure or timing error. The fourth level establishes the specific components, such as resistors, diodes, etc.

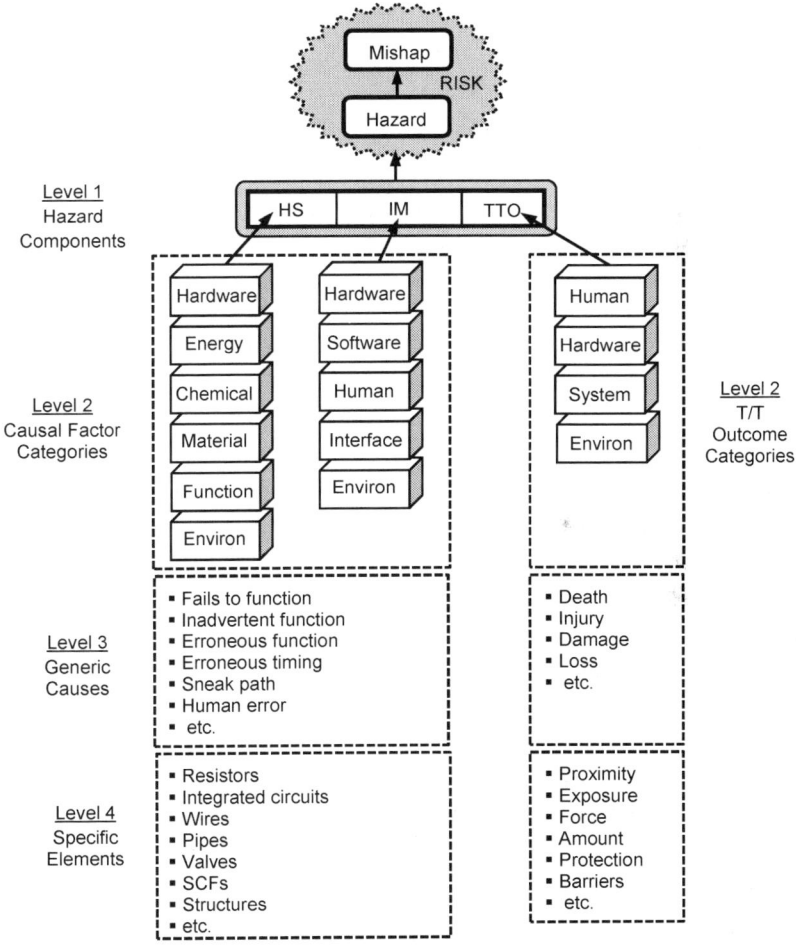

Figure 13.1 – Hazard Causal Factor Model

High level hazards in a PHA might identify root causes at the HCF category level, while a more detailed analysis, such as the SSHA, would identify the specific detailed causes at the component level, such as specific component failure modes. A hazard can be initially identified from the causal sources without knowing the specific detailed root causes. However, in order to determine the mishap risk and the hazard mitigation measures required, the specific detailed root causes must eventually be known.

In summary, the basic principles of Hazard-Mishap theory are as follows:

1) Hazards cause mishaps; a hazard is a condition that defines a possible future event (i.e., mishap).
2) A hazard and a mishap are two different states of the same phenomenon (before and after).
3) Each hazard/mishap has its own inherent and unique risk (probability and severity).
4) A hazard is an entity comprised of three components (HE, IMs, T/T).
5) The HE and IMs are the HCFs and they establish the mishap probability risk factor.
6) The T/T along with parts of the HE and IM establish the mishap severity risk factor.
7) HCFs can be characterized on three different levels.
8) The probability of a hazard existing is either 1 or 0; however, the probability of a mishap is a function of the specific HCFs.

A hazard is like a mini-system; it is a unique and discrete entity comprised of a unique set of HCFs and outcomes. A hazard defines the terms and conditions of a potential mishap; it is the wrapper containing the entire potential mishap description. The mishap that results is the product of the hazard components.

13.7 The H-4M Model

Risk management is the systematic application of management and engineering principles, criteria and tools to optimize all aspects of safety within the constraints of operational effectiveness, time, and cost throughout all operational phases. To apply the systematic risk management process, the composite of hardware, procedures, and people that accomplish the objective, must be viewed as a system. The basic causal factors for mishaps fall into the same categories as the contributors to successful operations, which are: Human, Media, Machine, Mission, and Management (H-4M).

The H-4M model, depicted in Figure 13.2, is adapted from military operations risk management (ORM). The H-4M's are Human, Machine,

Media, Management, and Mission. Human, Machine, and Media interact to produce a successful Mission (or an unsuccessful one). The amount of overlap or interaction among the individual components is a characteristic of each specific system and evolves as the system develops.

In this model, Human is used to indicate the human participation in the activity. Mission is the term that corresponds to purpose in function and duration of the system's operation. This model provides a framework for analyzing systems and determining the relationships between the elements that work together to perform the task. Management governs the interactions between the other elements by providing the procedures and rules. When an operation is unsuccessful or an accident occurs, the system must be analyzed; the inputs and interaction among the H-4Ms must be thoroughly reassessed. Management is often the controlling factor in operational success or failure.

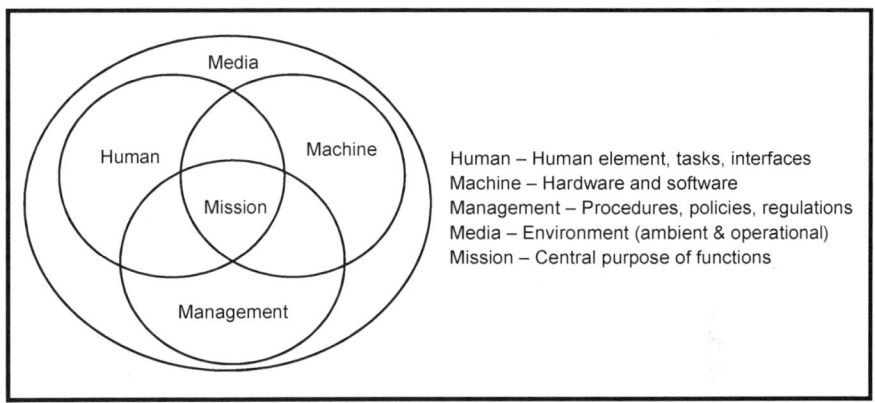

Figure 13.2 – The H-4M Model

The human factor is the area of greatest variability, and because of this variability, the human element has the possibility of creating many more risk situations than the other four factors. Some of the traits that must be examined when analyzing the human element in the safety risk equation include:

- Selection: Matching the right person psychologically and physically, trained in event proficiency, procedures and habit patterns to the task
- Performance: Awareness, perceptions, task saturation, distraction, channeled attention, stress, peer pressure, confidence, insight,

adaptive skills, pressure/workload, fatigue (physical, motivational, sleep deprivation, circadian rhythm) — are all factors that affect performance

- Personal Factors: Expectancies, job satisfaction, values, families/friends, command/control, perceived pressure (over tasking) and communication skills — are all factors that affect performance and adherence to safety requirements

Media are defined as external and largely environmental and operational conditions that affect operation and thus risk management. For example:

- Climatic: Ceiling, visibility, temperature, humidity, wind, precipitation
- Operational: Terrain, wildlife, vegetation, human-made obstructions, daylight, and darkness
- Hygienic: Ventilation/air quality, noise/vibration, dust, and contaminants
- Vehicular/Pedestrian: Pavement, gravel, dirt, ice, mud, dust, snow, sand, hills, curves

The machine is defined as the hardware and software comprising the system. Hardware and software even when used as intended, can pose risks to the operator, the environment, or to the equipment (machine) itself. The human-machine interface also adds to the risks accompanying any operation of a machine. Some of the areas where risk can be introduced into the machine are:

- Design: Engineering reliability and performance, ergonomics
- Maintenance: Availability of time, tools, and parts, ease of access
- Logistics: Supply, upkeep, and repair
- Technical data: Clear, accurate, useable, and available

Management directs the risk management process by defining standards, procedures, and controls for those elements, interfaces and interactions that are within its authority. Although management provides procedures and rules to govern interactions, management cannot completely control the system elements or, for the most part, the external inputs, such as the weather. Some management risk controls are:

- Standards: Policy and Orders
- Procedures: Checklists, work cards, and manuals

- Controls: Operator rest, restrictions, training rules/limitations and protocols

Mission is the overall stated purpose, goal or function of the system. That is, what the system was designed to do. Some of the risk factors affecting mission are:

- Objectives: Complexity understood, well defined, obtainable and well documented
- Interactions: Known and identified between and among Human, Media, Machine, and Management

13.8 Notes on Human Error HCFs

Human error is an important factor in designing for safety. Some human errors can be prevented, while it may not be possible to prevent other human errors. A HA should always consider situations where human error is safety-critical, and then mitigate these hazards by designing to protect against them. A HA should also consider poor designs that force a human to commit an error. There are several ways to categorize human error HCFs, such as:

- By fault type
    - Omission
    - Commission
    - Sequence error
    - Timing error
- By situation assessment versus response planning
    - Errors in problem detection
    - Errors in problem diagnosis
    - Errors in action planning and execution (for example: slips or errors of execution versus mistakes or errors of intention)
- By level of analysis
    - Perceptual (e.g., optical illusions)
    - Communication
    - Organizational
- By exogenous versus endogenous source (i.e., originating outside versus inside the individual)
- By physiology (burnout, depression, drugs, alcohol)

13.9 Notes on Organizational HCFs

Some safety literature considers organizational hazards as the source for most mishaps. This type of thinking is somewhat of a slippery slope, however, because organizational factors are typically not the root cause of a hazard or mishap. Organizational factors may be involved in the chain of events that allow a hazard to exist or persist, but not in the basic hazard root causal factors. The hazard root HCFs include factors such as failures, errors, sneak paths, excessive environments, common cause failures, etc.

Organizational factors are important because they are what can make or break a system safety program, and separate a poor safety design from a good one, or prevent unsafe operations.

13.10 Summary

This chapter has provided an overview on the various viewpoints regarding hazard causal factors. AN understanding of these perspectives will aid the analyst in the HA process.

CHAPTER 14

FAILURE DATA

14.1 Introduction

Many HA techniques require quantitative data in order to evaluate the hazard risk involved. HA quantification requires a probability of failure value for basic failure events. For example, a probability of failure (P_F) must be obtained for the failure event "relay fails open".

Quantitative failure data can come in the form of a frequency or a probability, depending on how it is obtained. Failure data can consist of actual field data or data estimates.

The question of where and how to obtain quantitative data for a HA is always a concern for safety analysts. This chapter focuses specifically on failure data and some resources for obtaining data.

14.2 Risk and Failure Rate

The metric of risk is utilized in many HAs. As covered in chapter 5, risk was explained as:

Risk = Likelihood x Severity

Likelihood refers to the likelihood of the hazard occurring and severity refers to the severity of the hazard outcome when it becomes a mishap. Likelihood can be evaluated in either qualitative or quantitative terms. Some HAs, such as the Subsystem Hazard Analysis (SSHA), utilize quantitative data for risk calculations. In this case, a probability is computed from identified component failures in the hazard causal factors.

If the failure data comes in the form of a probability, then no calculation is necessary and it can be directly used in the HA. If it comes in the form of a failure rate frequency, then a calculation is necessary to obtain a probability. Probability of failure is calculated using *failure data* that is directly related to the particular component, assembly or subsystem. The probability of a component failure is calculated using the reliability equation $P_F = 1 - e^{-\lambda T}$, where λ is the failure rate for the item and T is the exposure time for the item.

All components have one or more failure modes and failure rates associated with those modes. For example, a resistor can fail in the modes of: open, shorted or out of tolerance. Failure data can either be in the form of a prediction or historical field data. A predicted failure rate is one that has been predicted from various methods, such as analysis, test or prediction formulas. A predicted rate is used when there is no historical information or data on an item. A historical failure rate is one that has been derived from actual field use. As the item has more operational hours on it, the historical rate will approach its true failure rate value and is therefore the most useful.

14.3 Reliability and Failure Rate

Reliability is typically defined as "the probability that a device will perform its intended function, without failure, during a specified period of time under stated conditions". Reliability relies heavily on statistics, probability theory and reliability theory. The function of reliability engineering is to develop the reliability requirements for a product, establish an adequate reliability program, and perform appropriate analyses and tasks to ensure the product will meet its requirements. The failure rate of an item is a key value in reliability engineering.

In reliability theory, R is the probability of successful operation of an item and Q is the probability of unsuccessful operation (i.e., failure). Q is typically represented as P_F for probability of failure. The following equations are used extensively in reliability quantitative calculations:

- $R + Q = 1$
- $R = P_S = e^{-\lambda T}$
- $Q = P_F = 1 - e^{-\lambda T}$
- $\lambda = 1 / MTBF$ or $\lambda = 1 / MTTF$
- MTBF = (total hours of operation) / (total number of failures)
- Where λ is the failure rate and T is the exposure time.

Mean time between failures (MTBF) is a basic measure of reliability for *repairable items*. It is the mean number of life units during which all parts of the item perform within their specified limits, during a particular measurement interval under stated conditions. MTBF is the predicted elapsed time between inherent failures of a system, component or product during operation. MTBF can be calculated as the arithmetic mean (average) time between failures of an item. The MTBF is typically part of a model that assumes the failed item is immediately repaired (zero elapsed time), as a part of a renewal process. This is in contrast to the mean time to failure (MTTF), which measures average time between failures with the modeling assumption that the failed item is not repaired. Reliability increases as the MTBF increases. The MTBF is usually specified in hours, but can also be used with other units of measurement such as miles or cycles.

Mean time to failure (MTTF) is a basic measure of reliability for *non-repairable items*. It is the total number of life units of an item population divided by the number of failures within that population, during a particular measurement interval under stated conditions. MTTF measures the average time between failures with the modeling assumption that the failed item is not repaired. Reliability increases as the MTTF increases. The MTTF is usually specified in hours, but can also be used with other units of measurement such as miles or cycles.

14.4 Using Failure Data in a HA

In order to quantify a hazard, the probability of failure value is required for all basic failure events in the hazard. This value is typically derived from the failure rate for the item being investigated. Figure 14.1 demonstrates a quantitative evaluation for an example hazard. This example hazard evaluation demonstrates how component failure rates are used in a quantitative evaluation. The probability of failure of a component is a function of the component's failure rate and the component's exposure time during system operation. Hazard probability is obtained by multiplying the probability of each hazard component.

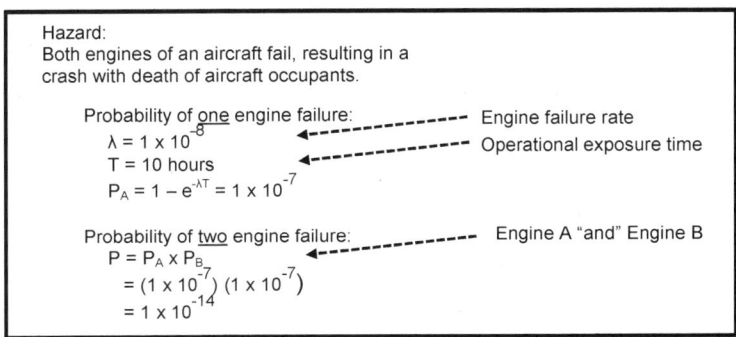

Figure 14.1 – Example Hazard Quantification

14.5 Obtaining Failure Data

Where do failure rates come from? This is a key question and concern of FTA analysts. Failure rate data can be obtained in several ways, the most common sources include:

A. Historical Data

Many organizations maintain internal databases of failure information on the devices or systems that they produce, which can be used to calculate failure rates for those devices or systems. For new devices or systems, the historical data for similar devices or systems can serve as a useful estimate.

B. Government and Commercial Data Sources

Handbooks of failure rate data for various components are available from government and commercial sources. Mil-Hdbk-217F, *Reliability Prediction of Electronic Equipment*, is a military standard that provides failure rate data for many military electronic components. Several failure rate data sources are available commercially that focus on commercial components, including some non-electronic components. Many government agencies maintain databases of historical data on the systems they operate.

C. Testing

The most accurate source of data is to test samples of the actual devices or systems in order to generate failure data. Companies developing products typically perform testing to determine the reliability of their product. Testing items produced by manufacturers is a means for obtaining data; however, this method is often prohibitively expensive or impractical.

D. Research

Individual research is a feasible but often difficult method for obtaining failure data. Researching magazines and news articles for reported failures and operational hours of items is sometimes productive. Obtaining manufacturer's warranty claims for their products also produces claimed failure rates.

E. Analyses

Quite often existing analyses may provide the data needed. For example, a Failure mode and Effects Analysis (FMEA) generally

Hazard Analysis Primer

contains failure data for components in the FMEA. If an FMEA has been performed on your system, it may be a good resource for failure data.

14.6 Commercial Failure Data Sources

Commercial data sources are available that provide excellent data. Some common sources for failure data include the following:

A. Electronic Components

- MIL-HDBK-217, Reliability Predictions of Electronic Equipment
- Telcordia SR-332 (Bell Laboratories Bellcore)
- PRISM (Alion System Reliability Center (SRC))
- RIAC 217Plus (Reliability Information Analysis Center)

B. Mechanical Components

- NSWC Standard 98/LE1, Handbook of Reliability Prediction Procedures for Mechanical Equipment, U.S. Naval Surface Warfare Center, September 30, 1998.
- WASH-1400 Reactor Safety Study, 1975.

C. Human Error Rates

- Gertman, David I. & Blackman, Harold S., Human Reliability & Safety Analysis Data Handbook, John Wiley & Sons Inc., New York, NY, 1984.
- WASH-1400, Reactor Safety Study, 1975.
- THERP – Technique for Human Error Rate Prediction is a methodology used in the field of human reliability assessment (HRA), for the purposes of evaluating the probability of a human error occurring throughout the completion of a specific task.

14.7 Failure Data Quality

Analysts are typically concerned about the quality of failure data. Quantitative results are only as accurate as the input data. Quite often it is difficult to obtain premium failure data with a proven history. When premium failure data is not available, worst case data can be used and still provide meaningful results. If the hazard probability is within acceptable ranges when worst case data is used, then it is not typically necessary to refine the data. If the probability is not acceptable, then refine the data only for the system elements causing the problem. When the results are not desirable, then, find more accurate data for those items of concern only.

Data uncertainty can be evaluated through the use of sensitivity analysis, research and testing. Failure data used in HA quantification sometimes contains an element of uncertainty for various reasons, such as:

- No data is available for a component
- Data is obtained from partial historical records
- Data is obtained from limited sampling and testing
- Data is obtained from prediction methods
- Data is obtained from manufacturer claims

14.8 Data Reality Check

A numerical probability of failure can go a long way toward confusing the risk management issue. How do you judge the acceptability of a numerical probability, for a given severity level? It often helps to compare it to other numbers that represent risk-like phenomena. The following are generalized risk calibration points that may help you relate to risk acceptance decisions (all are based on one hour of exposure)[10]:

- 10^{-2} – Human operator error in response to repetitive stimuli.
- 10^{-3} – Internal combustion engine failure (spark ignition).
- 10^{-4} – Pneumatic instrument recorder failure.
- 10^{-5} – Distribution transformer failure.
- 10^{-6} – Solid state semiconductor failure.
- 10^{-6} – Motor vehicle driver/passenger fatality.
- 10^{-6} – Lifetime average risk of death by disease.
- 10^{-7} – Flanged joint pipe blowout (4-in. pipe).
- 10^{-14} – Earth's destruction by collision with extraterrestrial body.

Remember, failure rates, exposure times and probabilities must be realistic and make sense in the real world.

14.9 Summary

This chapter has provided some useful considerations and resources for obtaining quantitative failure data and failure rates to be used in a HA.

---

[10] R. L. Browning, The Loss Rate Concept in Safety Engineering, Marcel Dekker, 1980.

CHAPTER 15

HAZARD ANALYSIS QUESTIONS

15.1 Introduction

In general HA seems like a relatively simple concept – analyze a system using a rigorous methodology and identify the hazards. Besides the confusion over the definition of a hazard, there are some basic complexities that must be considered while performing a HA. This chapter addresses the various issues, complexities and questions surrounding hazards, risk and HA.

15.2 HA Types vs. Techniques

Confusion often results from the concept of HA *types* and HA *techniques*. HA *type* defines an analysis category or class of analysis, whereas the HA *technique* defines a unique analysis methodology. In general, HA type defines the "what and when" to analyze for safety, while the HA technique defines the specific "how to" perform the analysis. Refer to chapter 10 for more detail on these two concepts.

15.3 Primary vs. Secondary HA

Quite often the question arises as to whether or not a particular HA technique qualifies as a *true* HA that can be depended upon for complete safety coverage. Within the system safety discipline there are over 100 different HA techniques that have been proposed, some of which are unique, some of which are merely variants of others, some of which are extremely useful and some of which are not useful at all. Some of the proposed HA techniques are not true HAs. Essentially, there are *primary* and *secondary* HA techniques; secondary techniques must not be solely relied upon for system safety. Refer to chapter 10 for an explanation of which techniques are primary and secondary.

15.4 HA Tailoring of Techniques

A common HA question involves HA tailoring on a system safety program. MIL-STD-882 recommends that all seven primary HA techniques of: PHL, PHA, SSHA, SHA, O&SH, HHA and SRCA always be performed on each system development program, when feasible. However, MIL-STD-882 also allows HA tailoring when performing all seven techniques does not seem appropriate or feasible. Depending upon the size, cost, complexity and criticality of the system, the specific HA techniques to be applied can be tailored to the program. The selection of HAs to be used should be selected judiciously and the rationale documented. Table 15.1 provides some minimal guidelines that I recommend for HA tailoring.

Table 15.1 – HA Tailoring Guidelines

Small Dollar or Low Risk System	Medium Dollar or Average Risk System	Large Dollar or High Risk System
PHL FHA PHA SHA SAR	PHL FHA PHA SSHA SHA O&SHA HHA SAR	PHL FHA PHA SSHA SHA O&SHA HHA SRCA FTA SAR
PHL - Preliminary Hazard List analysis PHA - Preliminary Hazard Analysis SSHA - Subsystem Hazard Analysis SHA - System Hazard Analysis O&SHA - Operations and Support Hazard Analysis HHA - Health Hazard Assessment FHA - Functional Hazard Analysis SRCA – Safety Requirements/Criteria Analysis SAR - Safety Assessment Report		

15.5 Minimum HA Techniques Coverage

How many HAs should be performed on a typical system? Typically, multiple HA methodologies are applied to a system to ensure complete system coverage and complete hazard identification. This is necessary to ensure complete safety coverage of the primary system viewpoints mentioned earlier (chapter 7): physical, functional, operational, software, human, environment and organizational. This is not a hard and fast rule, but depends upon various factors, such as: system size, system complexity, safety criticality of the system, etc. The number and type of HAs to be performed should be planned in advance and tailored to the specific system and the system requirements involved.

There are only a handful of HA techniques that are commonly used by system safety experts. When possible it is advisable to perform all of the techniques shown in Table 15.1 under "average risk system". However, when this is not feasible I recommend that the following HA techniques always be performed as a bare minimum: PHL, FHA, PHA and SHA. Refer to chapter 10 for a brief discussion of these techniques.

15.6 One HA or Multiple HAs?

Another HA confusion factor is in regard to the question of how many HAs should be performed on a system during system development. Is only one HA necessary or are several different HA techniques required?

This is somewhat of a mixed question. On the one hand, MIL-STD-882 established the seven HA types that should be performed to ensure all hazard types are discovered. On the other hand, perhaps a single HA technique would suffice if it were iteratively updated as the design progressed. This would require an experienced analyst that knew how to identify all types of hazards to make it fit into one HA.

It is also a matter of system complexity, system safety criticality and what is required by the customer. Based on experience, I recommend that as a minimum, the PHL, FHA, PHA and SHA are always performed. A safety assessment report (SAR) summarizing the hazards and system risk can also be included. See table 15.1 above for additional guidance information.

15.7 One Hazard or Two?

Quite often a hazard will be identified that has several different and unique causal factors, but the same overall undesired outcome. In this situation, each of the unique set of causal factors will cause the same general outcome. Some safety analysts see this as one hazard with several causal factors, while other analysts see this as several hazards, each having the same outcome.

To illustrate this concern, a simple missile launch system example is provided for HA, as shown in Figure 15-1. The missile launch function is executed when the Missile Launch electronics receives the +28 VDC missile launch signal. There are two interlocks for the launch signal, Relay-1 and Relay-2, which must be closed in order to send the launch command.

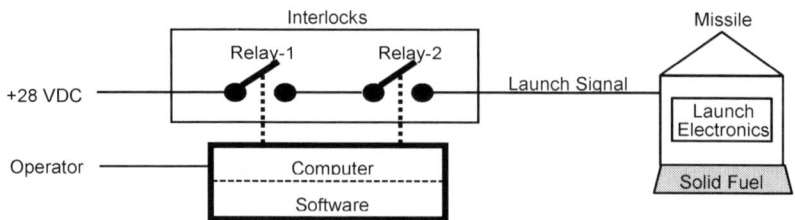

Figure 15.1 – Example Missile Launch System

The hazard in this example involves "Inadvertent Missile Launch". Analysts in Group A would categorize this as one hazard, stated as:

Hazard #A1
"Inadvertent Missile Launch caused by hardware faults and/or software faults".

Analysts in Group B would see this as several hazards with a common outcome, stated as:

Hazard #B1
Inadvertent Missile Launch caused by spontaneous ignition of the solid fuel motor. Note that this is a hardware casual factor.

Hazard #B2
Inadvertent Missile Launch caused by component failures in Launch Electronics unit. Note that these are hardware casual factors.

Hazard #B3
Inadvertent Missile Launch caused by Relay-1 failing closed AND Relay-2 failing closed. Note that these are hardware casual factors.

Hazard #B4
Inadvertent Missile Launch caused by Relay-1 AND Relay-2 being inadvertently closed by computer hardware failures. Note that these are hardware casual factors.

Hazard #B5
Inadvertent Missile Launch caused by Relay-1 AND Relay-2 being inadvertently closed by software errors. Note that these are software casual factors; the software controls hardware to formulate the hazard.

In order to accurately identify the risk involved with this hazard, it must be broken into five unique hazards, with each hazard presenting its own level of risk. If this were one hazard, the true risk would be masked. Remember, "Resistor R45 failing open, resulting in missile launch causing death/injury" and "Switch S7 fails closed, resulting in missile launch causing death/injury" are two hazards, not one hazard with two different causal factors. The hazards are related because they each have the same overall undesired outcome. Also, the hazard written by group A violates the hazard rule of "hazard causal factors can only be ANDed together, and OR situation (generally) indicates multiple hazards".

15.8 Hazard and Risk Confusion

In addition to the confusion over one hazard vs. several (see above), there is a related question of how hazard causal factors work together and where the risk is calculated. Some of the beliefs associated with this viewpoint include the following:

- A mishap outcome can be caused by more than one hazard or by one of several causal factors triggering the single hazard that causes the mishap.

- Just as multiple hazards can lead to the same mishap, multiple causal factors can trigger the same hazard, which then causes a mishap.

- Risk is only assigned to hazard causal factors.

These erroneous statements are shown graphically by the diagram shown in Figure 15.2. There is a scant bit of truth in these statements, but they are not organized in the proper fashion to present the complete and correct story.

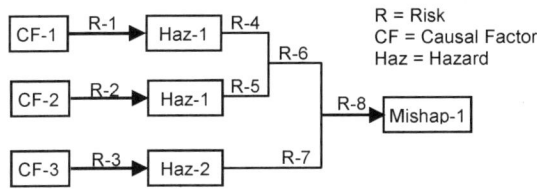

Figure 15.2 – Erroneous Hazard-Mishap Relationships

The model shown in Figure 15.2 violates the basic laws of hazards-mishaps relationships. The correct hazard-mishap relationships are shown in Figure 15.3, which correctly depicts the basic laws of hazards-mishap relationships.

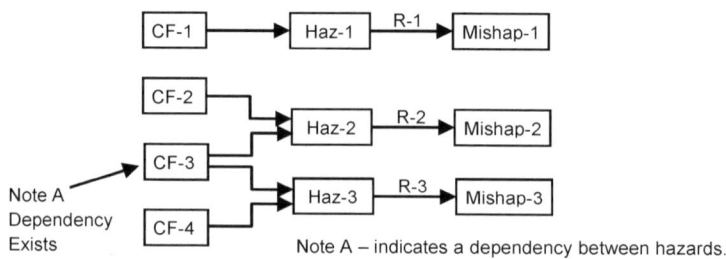

Figure 15.3 – Correct Hazard-Mishap Relationships

The hazard-mishap relationship conclusions to be drawn from the errors in Figure 15.2, and correctly modeled in Figure 15.3, include the following:

a) Risk is a function of both hazard likelihood and severity. This information can only be derived from the hazard description, which includes both causal factors and severity factors. Thus, risk items R-1 through R-8 in Figure 15.2 make little sense.

b) Risk must be mathematically combined at the hazard level.

c) When the same causal factors support two different hazards, then there is an element of dependence involved. This means the two hazards are not independent and more complicated mathematics is involved.

d) In general, one hazards results in only one mishap. Several hazards can contribute to the same generic TLM outcome, but not the same exact mishap.

These two examples illustrate the hazard rule of "One hazard ... one mishap". Multiple causal factors to the same general outcome cause confusion for many analysts.

15.9 Hazard Count; Too Many Hazards?

Hazard count often causes confusion and/or concern. Quite often a large and complex system can have 2,000 hazards, or more. This sometimes concerns system developers because they (erroneously) think this means their system is unsafe. The number of hazards is not what makes an unsafe system; it is the residual risk presented by the hazards after HA and mitigation that reflect the level of safety.

Sometimes system developers will try to bundle several hazards into one hazard, in order to reduce the count. This is not recommended because it hides hazard and risk visibility. It also violates the general hazard rule of hazard causal factors can only be ANDed together (an OR situation generally indicates multiple hazards). In reality, the hazards are being combined under a generic TLM outcome and not a single hazard.

15.10 Hazard Risk Summing

Quite often system developers, evaluators, testers and/or the operators desire to know the total system risk (TSR) of the system. TSR is (theoretically) the sum total potential mishap risk that is inherent within the system, which would equate to the total risk presented by all system hazards in the system. TSR is the total mishap risk for a system; in theory it is the sum total risk for all hazards in a system. It's essentially a theoretical value.

Wouldn't it be useful to know the TSR for an existing system, or a system under development? A TSR metric for a system would make it very easy for program managers and system users to make decisions regarding the relative safety level of a system, and to compare the risk presented by competing systems or alternative design methods. Risk summing is an important topic in HA and system safety. It would be nice to have a single number that told the program manager how much safety risk his/her system presented. This would alleviate critical thinking and decision making.

Although TSR would seemingly be an extremely useful metric, the question is.....can TSR be realistically computed? The short answer is NO; in reality it's a theoretical concept and an elusive goal. The long answer is that it may be possible, but it's a difficult and error prone process to compute a TSR value, and therefore not really cost-effective. In addition, it might give a false sense of security if any hazards are unidentified and left out of the equation. However, other mishap risk metrics and methods, such as TLM risk, are easier to compute and are very useful, if not more useful.

The following are some reasons explaining why it's not possible or useful to obtain or compute the TSR for a system:

- Safety risk is obtained from hazards, which means that it is necessary to identify *all* system hazards in order to obtain *all* risk (i.e., TSR). Although safety analysts do their best to identify all hazards, we all know that it's a difficult and error prone process, and often some hazards are overlooked. Overlooked hazards would under value the computed TSR.

- Mishap risk involves the components of probability and severity, and it's not a simple process to mix severity levels in a total composite number; it's like mixing apples and oranges (e.g., death + minor injury = ?). Severity levels would have to be converted to a common denominator, such as dollar loss, in oder to be summed.

- It's not possible to sum hazard risks that are computed qualitatively. The hazard and risk indices in MIL-STD-882 are for individual hazards. There is no guidance as to the result when two indexed values are summed. For example, what is the result of a 1C hazard added to a 2D hazard?

- In order to try to mix severity categories a consistent quantitative baseline is necessary, which means that a PRA is required, and quantitative values must be established for losses such as death, injury, environmental damage, etc. In a cost effective program performing PRA on low risk hazards is generally cost prohibitive and ineffectual, yet a TSR cannot be obtained without them.

- Risk numbers deal with probabilities as one component, and probabilities can introduce a large error margin in a single hazard risk calculation, thereby skewing the results of a TSR calculation. Competing companies might use favorable probabilities in order to make their product look better, resulting in a questionable TSR value.

- In a typical system safety program (SSP), generic safety requirements are applied to mitigate a certain category or class of hazards. For example, MIL-HDBK-454 and MIL-STD-1472 are generically applied for electrical, mechanical and temperature hazards to personnel as a generic requirement. As a result, not all of these hazards are identified and summed in a hazard analysis because the generic requirements cover them. This mishap risk would be missing from a TSR assessment.

- In a typical SSP not all hazards and hazard causal factors are known until late in the final design process. This means that an accurate TSR value could not be obtained early in the program. This also means that agencies comparing bid proposals could not effectively use any kind of TSR to compare competitor bid proposals.

- The method of counting hazards is a critical issue. For example, some analysts perceive one hazard with three sets of causal factors, whereas other analysts see three discrete hazards for the same situation. The method of accounting could make a big difference in the final TSR value.

Hazard Analysis Primer

- Dependence between the causal factors of various hazards will require special mathematical summing methods in order to avoid mathematical errors between dependent events. In addition, dependence may not be obvious without significant analysis to identify them.
- A calculated TSR would have to be used very cautiously because of the possible variance in an accurate result. The value might provide a false sense of security on the one hand, or it may provide a feeling of panic on the other.
- Some hazards cannot happen if certain other ones do happen, i.e., they are mutually exclusive. A TSR calculation would necessitate identifying these cases and applying special mathematics. For example, inadvertent launch would preclude inadvertent ground detonation.
- Hazard risk is associated with a potential future undesired event. To sum the risk of many hazards is to attempt to combine many potential future events, many of which have different severity levels. Risk probabilities can be summed (if they are independent) but summing severity levels is very tricky and may result in meaningless results.
- Looking at the investment industry, analysts may study four alternatives investment ideas, each with its own risk. These analysts compare the risk, but they don't sum them. It would make no sense to sum them.

Total system risk is a nice concept; however, deriving a TSR value is a theoretical perception. For the above mentioned reasons, it is not cost-effective to derive a TSR value. In addition, the process is error prone, which would very likely make any final TSR incorrect, resulting in a false confidence in safety provided by the results. What does a total risk sum really mean and how would it be used to justify the expense? Obtaining and correctly using a TSR would probably be too risky!

Useful system risk metrics can be more effectively obtained from summing hazard risks within a TLM category. This value would be more of a total mishap risk (TMR), because it sums al of the risks for a particular mishap. This is more feasible because each mishap, or TLM, would have the same severity level. A system has many different and diverse TLMs having different outcomes and severities. Since these outcomes are totally different, they cannot be directly summed in terms of probability and risk. Just as different fault trees with different top undesired events cannot be summed. Only the events under a single fault tree TUE can be summed together.

15.11 Risk Assessment Tailoring

Another complexity regarding HA and system safety involves the tailoring of risk tables. Should risk acceptance criteria be standard for all systems or should the criteria be tailored to fit different systems and system types? MIL-STD-882 provides a suggested risk matrix (see section 5.8) for evaluating the safety criticality of hazards; it also states that the matrix can be modified or tailored to fit the needs of a program.

If tailoring of these matrices is desired, there are several aspects that can be modified, such as:

- The severity category definitions in Table 5.1 can be modified
- The probability category definitions in Table 5.2 can be modified
- The risk index cells definitions in Table 5.3 can be moved around
- The risk level index value definitions in Table 5.4 can be modified

The question that must be answered by each individual program is: what makes sense to change? Risk can easily be under-valued or over-valued if the wrong decisions are made. Also, can different programs really be effectively compared if they each use different risk tables?

15.12 Qualitative vs. Quantitative Risk Evaluation

A question that is often asked is: which is better, qualitative or quantitative risk assessment? This is a complex question because the answer can vary. To demonstrate this complexity, two example scenarios are considered. In example #1 qualitative assessment has the advantage, while in example # quantitative analysis has the advantage.

In example #1 three hypothetical hazards are rated quantitatively and qualitatively, as shown in Table 15.2. Each hazard has a different probability of occurrence and a different dollar loss value. However, note that the resulting quantitative risk value for each hazard is identical, that is, they each have a quantitative risk value of 10. Based on this information, which hazard is more critical and should receive mitigation first? The quantitative risk value does not indicate which hazard is more significant.

Table 15.2 – Risk Example Comparison #1

Hazard	Quantitative			Qualitative			
	Loss (L)	Probability	Risk (L x P)	Severity	Probability	Risk (S x P)	Risk Index
#1	$10 M	1×10^{-6}	10	I	E	1E	12
#2	$100 K	1×10^{-4}	10	III	D	3D	14
#3	$100	1×10^{-1}	10	IV	A	4A	13

⇧ No Granularity   ⇧ Granularity

Using the qualitative definitions from a standard HRI matrix, the three hazards are assigned qualitative severity, probability and index values as shown in Table 15.3. Now it is very easy to see a distinguishing difference in risk levels between the three hazards; it provides more levels of distinction between the hazards. Hazard #1 fits into risk index #12 because it was assigned a 1E based on fitting into the appropriate risk definitions. It presents the most risk and should be mitigated first. Hazard #3 fits into risk index #13 because it was assigned a 4A based on the appropriate risk definitions. Although the loss value in dollars for Hazard #3 is much less than Hazard #2, its probability of occurrence is much higher, therefore making it a higher risk than Hazard #2. Hazard #2 presents the least risk and should be mitigated last, although its loss dollar value is much higher than Hazard #3. The small dollar loss value for Hazard #3 could be grounds for ignoring the risk ranking and allocating of resources to mitigate Hazard #2 before Hazard #3.

Table 15.3 – Risk Index for Hypothetical Hazards

Probability	Severity			
	I Catastrophic > $1M	II Critical $200K – 1M	III Marginal $10K – 200K	IV Negligible $2k – 10K
A) Frequent $< 10^{-1}$	1	3	7	13 **Haz #3**
B) Probable $10^{-1} - 10^{-2}$	2	5	9	16
C) Occasional $10^{-2} - 10^{-3}$	4	6	11	18
D) Remote $10^{-3} - 10^{-6}$	8	10	14 **Haz #2**	19
E) Improbable $> 10^{-6}$	12 **Haz #1**	15	17	20

In example #2 four hypothetical hazards are rated quantitatively and qualitatively, as shown in Table 15.4. Each hazard has a different probability of occurrence and a different dollar loss value. In this case the resulting quantitative risk values range in value from 20 to 1960. Based on this quantitative risk information, it is fairly obvious that Hazard #1 presents the highest risk, and Hazard #4 presents the least risk. The quantitative risk value does not indicate which hazard is more significant.

In this example, the quantitative method required much time and money to determine the correct values, and in the end, the risk values for the three hazards where indistinguishable by the quantitative method. The qualitative method did not require detailed probability calculations, but worked quite effectively with estimated ranges. Also, the qualitative method provided a means for distinguishing between the risk levels for each of the three hazards. In this case the qualitative approach appears to be more effective than the quantitative approach.

Table 15.4 – Risk Example Comparison #2

Hazard	Quantitative			Qualitative			
	Loss (L)	Probability	Risk (L x P)	Severity	Probability	Risk (S x P)	Risk Index
#1	$198 K	0.99×10^{-2}	1960	III	C	3C	11
#2	$198 K	1×10^{-3}	198	III	C	3C	11
#3	$20 K	1×10^{-2}	200	III	C	3C	11
#4	$20 K	1×10^{-3}	20	III	C	3C	11

Granularity   No Granularity

It can be deduced from this example, that when all of the hazards lie within the same qualitative probability and severity range, the qualitative method does not provide as much distinction between the hazards for ranking. In this case the quantitative approach provided a little more capability for ranking hazards by risk. However, it should be noted that this does not make the qualitative approach less valuable. On the contrary, it is still a very effective approach because it allows the analyst to quickly establish the relative risk of each hazard, and place them in the appropriate risk index category. This group of hazards will then be rated against other hazards with different risk index values.

15.13 Are Equal Hazard Risks Truly Equal?

An often asked question is in regard to equal risk hazards. If several hazards all present the same risk, are they equally dangerous? Figure 15.4 contains three hazards that contain equal risk. Although each hazard presents equal risk (area inside the rectangle), the probability and severity levels for each hazard are significantly different. For example, Hazard-1 (Haz1) has an extremely small probability but a very large severity level and Hazard-3 (Haz3) has a low severity level, but a high probability. Because risk is the product of probability and severity, these three hazards result in the same level of risk.

Since the risk is identical for all three hazards, does this mean that all three hazards are equally acceptable, or unacceptable? Does the probability-severity product portray an accurate picture of risk, or is there more to consider? Hazard Haz1 will likely receive greater focus because its severity level is high (catastrophic), even though the probability is small. Regardless of equal risk, catastrophic hazards must be judiciously compared and judged with low severity hazards.

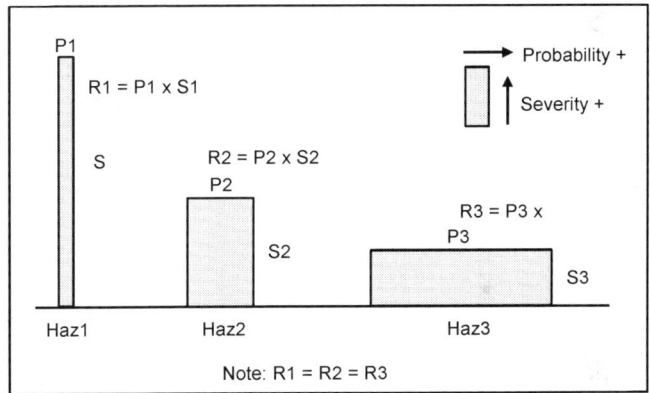

Figure 15.4 – Equal Risk Hazards

This example demonstrates one reason why the HRI matrix is useful for risk acceptance decisions. When hazard risks fall into buckets it makes acceptance decision making easier and removes the *dread factor* from hazards that have a high severity level (i.e., catastrophic).

15.14 System Risk Picture

At some level in the system hierarchy the risk presented by hazards must be summed. Risk aggregation is the roll-up of low-level risks to higher levels in order to obtain an overall system risk picture. This technique normally relies on qualitative judgment and team-voting methods to summarize risks at the critical risk area and process level in terms of likelihood and consequence.

The purpose of system risk aggregation is to give a more complete picture of "system" risk. A major concern is that a single top number may take-away from individual hazard detail. Another concern is that risk cannot be validly summed for reasons such as:

- Dependency between hazards affects the mathematics involved
- Hazard severity levels are significantly different
- Families of hazards are ignored in favor of one representative worst-case hazard

Figure 15.5 shows relative hazard probability levels at various system levels. This is a generic diagram depicting that component failure probabilities start at the bottom of the system and gradually roll-up by system level. At the top system level risk can be expected to be greater than at the bottom level. This diagram leads into the next section on where to score risk. Generally speaking, the HRI matrix is geared for hazards just below the major subsystem level.

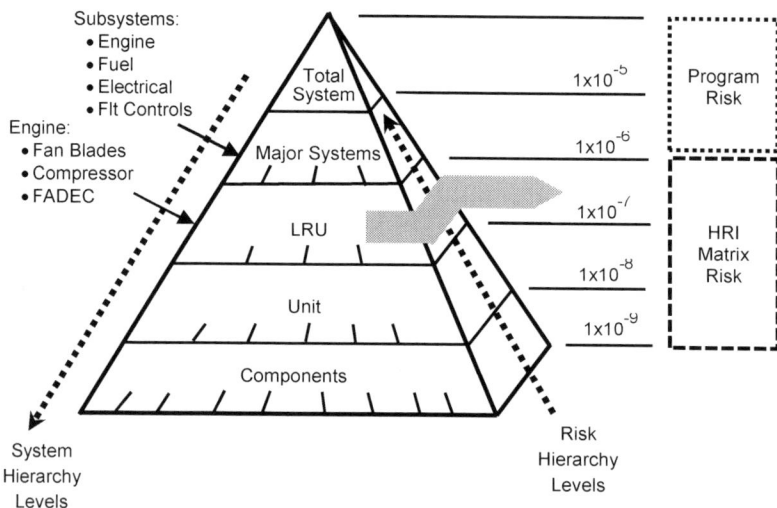

Figure 15.5 – Risk Levels

15.15 At What Level Should Hazard Risk Be Scored?

One of the interesting (and confusing) phenomena of hazard risk is that hazard risk can actually be "scored" at different levels in the system hierarchy. Scoring refers to documenting and accepting the risk. Risk can be scored at the component level, the assembly level, the unit level, the subsystem level or the system level. Each level brings its own advantages, disadvantages and problems.

This phenomenon generates an area of misunderstanding and confusion in HA, particularly in regard as to where to score hazard risk in the system hierarchy. Hazards are identified at the component level, assembly level, subsystem level and system level. Can risks at the assembly level be compared to risks at the system level? Can hazards (or risks) be rolled-up or rolled-down?

These questions involve the fact that hazard risks scored at the wrong system hierarchy level can hide and mask risk criticality and judgment. It is not safe to generalize that risk should be scored at a certain level, because the rationale can change. System indenture level can make a difference in a risk assessment.

When performing hazard analysis, all of the system components must be considered to ensure a complete analysis. A hierarchy table aids the safety analyst in ensuring that all of the system hardware and functions have been adequately covered by the hazard analyses. System hierarchy level can be used to set the level of detail for a particular hazard analysis.

A common question asked by safety analysts: at what level should a hazard be scored for an HRI; the system, subsystem, component or function level? If the HRI is rated at the system level, then many causal factors are covered up, and their individual risk significance is not visible. This does not allow prioritizing hazards and fixing the most significant ones. Also, the HRI matrix is not intended for high level system hazards.

If HRIs are scored at the component level, then the low failure rate of an individual component will yield a low HRI value. This would mask the true hazard risk in cases where the components must be summed at the LRU or subsystem level, and the HRI computed at that level.

Figure 15.6 examines the risk of a subsystem when scored at different system levels. This example involves a system with one Flight Control Computer (FCC) subsystem. The hazard is Loss of Function (LOF), which results in loss of aircraft (LOA). Assume for simplicity that each component failure will result in loss of FCC and LOA.

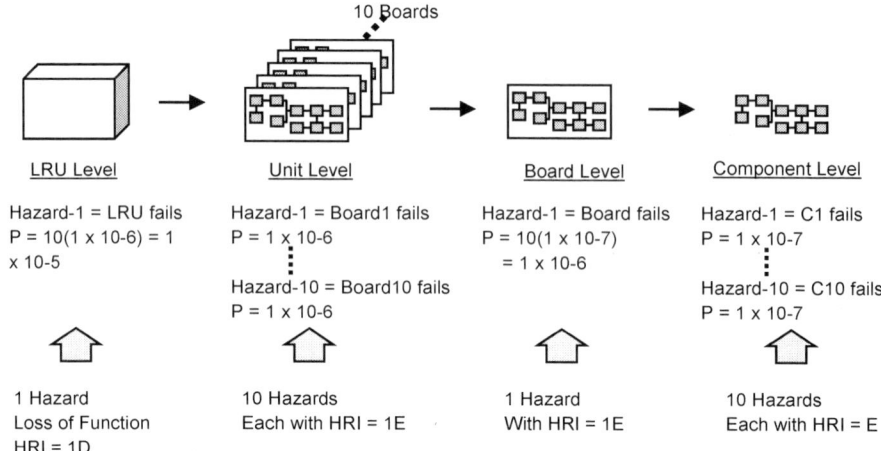

Figure 15.6 – Hazard Risk Levels: Case A

Figure 15.7 examines the risk of a subsystem at the subsystem level when redundant subsystems are involved. This example system involves three redundant FCCs, where only one is needed for system success.

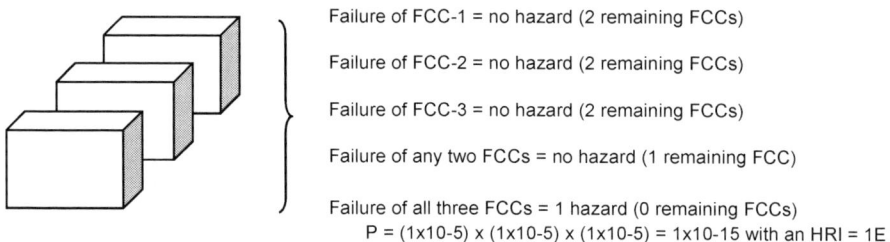

Figure 15.7 – Hazard Risk Levels: Case B

Some conclusions that can be drawn from case A and case B include the following:
- Need to determine hazard risk level on a case by case basis
- System architecture is a key factor
- System functions and LOF is a key factor
- Diversity of causal factors (across several LRUs vs. all in a single LRU)
- LOF hazard vs. malfunction (e.g., inadvertent/erroneous) type hazard

One issue in HA is determining at exactly what level to score hazard risk in the overall hierarchy of a hazard. Figure 15.8 shows the hierarchy breakdown of an aircraft propulsion system hazard using a fault tree (FT) diagram. One hazard, labeled Haz-1 is "engine fan blade fails causes loss of engine which results in loss of aircraft propulsion and loss of aircraft". The majority of this hazard is shown via FT gates G1, G2 and G3. The faults under G3 are the hazard causal factors (HCFs).

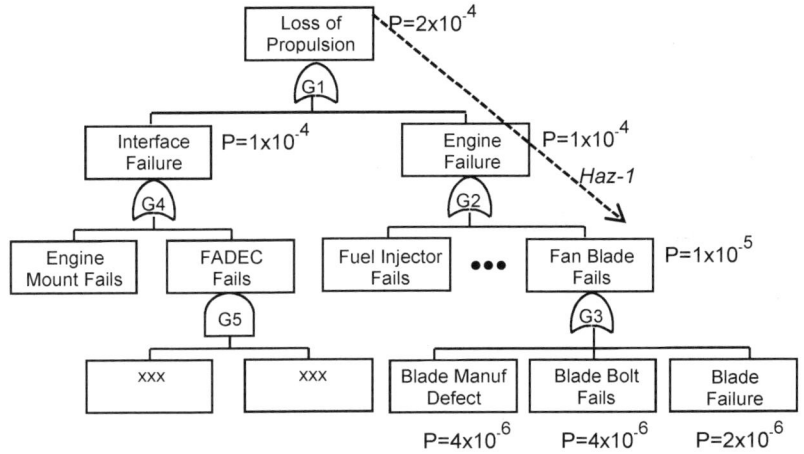

Figure 15.8 – FT of Aircraft Propulsion Hazards

This FT diagram shows the hazard risk probability component at different levels in the hazard hierarchy. At the gate G3 level the HCFs are summed (ORed) to yield a probability of P=1 x 10-5. Using the HRI matrix, the risk would be scored as 1D (catastrophic severity and remote probability P=10-5). The risk for this hazard would have to be accepted by someone high in the chain of authority.

Figure 15.9 shows Haz-1 being broken down into three hazards following the general rule that hazard HCFs should not be ORed. Note that at this level the hazards would each be scored at the 1E risk level (P=10-6). At this level the risk for each hazard could be accepted by the program manager. An advantage of scoring the risk at this level is that the hazard's risk can be accepted in a much easier process. However, at this level the overall safety-criticality of the fan blade is somewhat obscured.

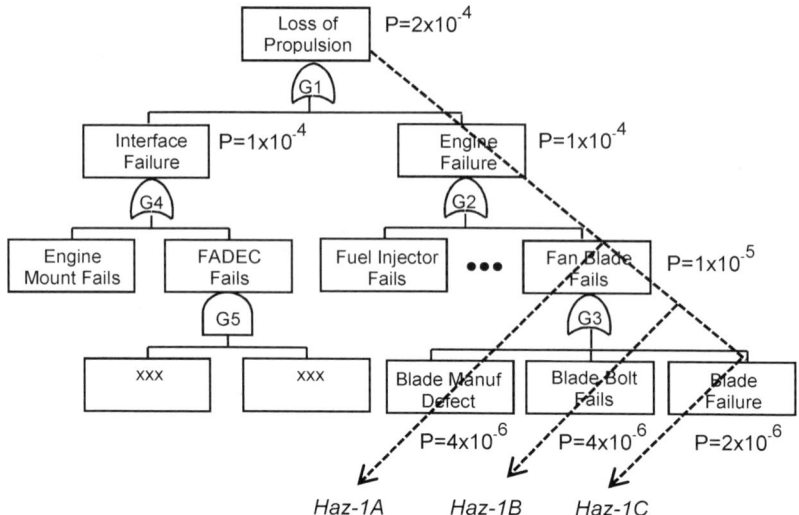

Figure 15.9 – FT of Aircraft Propulsion Fan Blade Hazards

There are obvious advantages and disadvantages in scoring risk at different levels. The question is what makes the best sense for the program and for enhancing safety. Should who accepts the risk drive the risk level? Should simplifying the HA process be the driver? These questions must be answered prior to the HA and documented in the HA plan.

15.16 Single or Multi-Thread Analysis

A question that often arises from program engineers is whether it's really necessary to perform a HA that covers dual failures. Another related question is, if you are evaluating redundant items how should the failure of a single item be treated in the HA, since it seldom results in a hazard.

Consider the small example system shown in Figure 15.10. This system is comprised of two sets of redundant hardware elements that provide output power. The generators are controlled by common software.

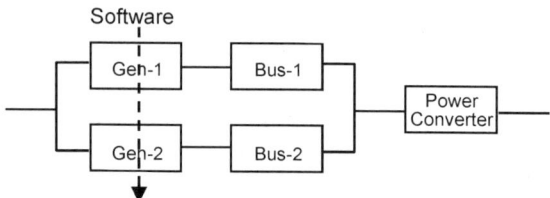

Figure 15.10 – Power Generation System

Hazard Analysis Primer

Table 15.5 contains an abbreviated HA of the electrical power generation system. In this single item HA approach, each item in the system is evaluated for the effect of its failure only. This is similar to the single thread analysis in an FMEA. Note that table items 1 through 4 result in no hazard. This is because the items are redundant and their redundant counterpart will take over and provide safe operation.

Table 15.5 – Single Item Approach

Hazard : Loss of all power results in loss of air vehicle control and crash (HRI = 1E)		
	Causal Factor (CF) Description	Comments
1	Gen-1 Fails	Gen-2 still available – no hazard
2	Gen-2 Fails	Gen-1 still available – no hazard
3	Bus-1 Fails	Bus-2 still available – no hazard
4	Bus-2 Fails	Bus-2 still available – no hazard
5	Power Converter Fails	No power output – hazard
6	Gen software error	Causes both Gen to fail – hazard

Table 15.6 contains another abbreviated HA of the electrical power generation system. In this multi-thread HA approach, only specific failures that can actually cause the hazard are considered. This is similar to the multi-thread analysis in a FTA. Note that in this approach all table items are hazards. This is because dual failures are considered that will cause failure of intended redundancy. Even though the likelihood of dual failures is typically smaller than single items, it is necessary to consider them and determine the actual risk presented. Sometimes dual failures are more likely than a single failure, especially if common cause failures are considered.

Table 15.6 – Multi-Thread Approach

Hazard : Loss of all power results in loss of air vehicle control and crash (HRI = 1E)		
	Causal Factor (CF) Description	Comments
1	Gen-1 and Gen-2 Both Fail	No power available – hazard
2	Bus-1 and Bus-2 Both Fail	No power available – hazard
3	Gen-1 and Bus-2 Both Fail	No power available – hazard
4	Gen-2 and Bus-1 Both Fail	No power available – hazard
5	Power Converter Fails	No power output – hazard
6	Gen software error	Causes both Gen to fail – hazard

The single item approach considers the item over the hazard. When redundancy is involved there is typically no resulting hazard. Is this approach a waste of time? No its not, because it does reveal hazards resulting from single failures. The problem is, how is the postulated hazard recorded and scored in the HA when no hazard is involved.

The multi-thread approach considers the hazard over the item. When redundancy is involved the HA considers what failures are necessary to defeat the redundancy. Is this approach a waste of time? No its not, because it does reveal safety-critical hazards resulting from dual failures. There is no question on how to record and score these hazards.

The bottom line is that both methods of analysis are necessary in order to ensure that all hazards are identified. Also, it is not just import, but necessary, to consider and evaluate dual failures in order to properly address safety.

15.17 Using or Abusing the Hazard-Risk Hierarchy?

The question of where in the system hierarchy and the hazard hierarchy to score risk is always an important one. It is very easy to abuse risk levels in order to beat the risk assessment process, where higher risks must be accepted by a higher level authority. If the risk can be driven down to a lower level, then a lower level authority can accept the risk. This is a trick I have actually seen used to get risk into an acceptable region. Which level makes the best sense to collect risk and apply risk acceptance?

Figure 15.11 depicts an example aircraft system that is broken down into several levels in the system physical hierarchy. The following discussion is only valid for one common hazard – loss of function causing aircraft crash. A dissimilar hazard, such as fire, could not be handled this way. The following are conclusions that can be drawn from each level:

- Level A represents one hazard risk caused by all of the aircraft Subsystems (the sum of the subsystems adds to $P=1\times10^{-4}$).
- Level B represents hazards caused by each individual Subsystem. For the Engine Subsystem only the probability is $P=1\times10^{-5}$ (the sum of all the Units within the Engine Subsystem).
- Level C represents hazards caused by each individual Unit within the Engine Subsystem. For the Engine Subsystem Unit A only the probability is $P=1\times10^{-6}$ (the sum of all the Assemblies within Engine Unit A).
- Level D represents hazards caused by each individual Assembly within the Engine Unit A. For the Engine Subsystem Unit A

Assembly A only the probability is P=1x10-7 (the sum of all the Boards within Assembly A).

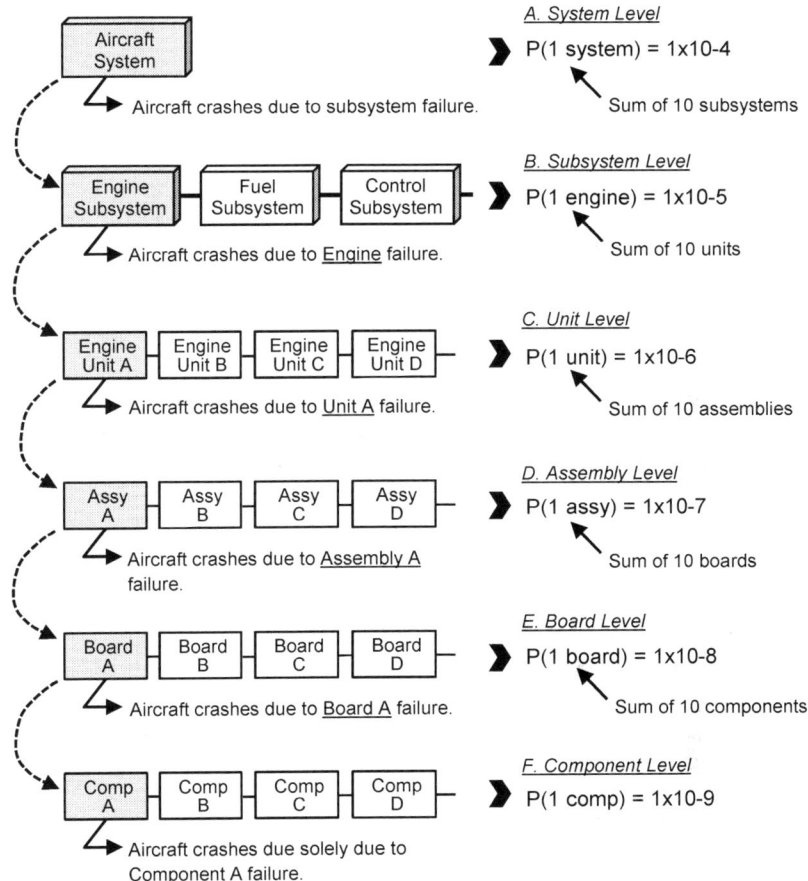

Figure 15.11 – Aircraft System Hierarchy

If the risk is scored at the Subsystem or Unit level it may require a very high person in the chain of command to accept the risk. If the risk is carried down to the Component level if may only require the program manger to accept the risk. Does this enhance the process or bury the risk? On the other hand, it's typically better to track and mitigate hazards at the Unit or Assembly level. Note that this is a very rough example, and that each item at one level is comprised of 10 items in the next lower level. This accounts for the probability changing by one order of magnitude at each level.

Trying to set hard and fast rules as to where to score hazard risk is a difficult task, and there always seems to be exceptions after making the rules. Selecting the right system level for risk acceptance is not simple or straight forward and there is no simple rule. There tends to be a natural tendency to try and write and score hazards at the subsystem or unit level. Sometimes this works well and sometimes not.

15.18 Families of Hazards

Do systems present families of hazards? Families of hazards involve the idea that if one parameter in a hazard's causal factors can vary, then each possible variation presents a slightly different outcome and thus a family of hazards. For example, if a worker is working with rotating machinery and a part flies off hitting him, he could be hit in the finger, arm, shoulder or head. Depending on where he was hit the injury would be more severe. It would seem that each hazard in the family would add to the total risk, however, most HAs only consider the worst case outcome.

15.19 Summary

This chapter has presented and discussed some of the pressing questions surrounding the HA process. These questions demonstrate the complexity of HA and the fact that it is not a trivial process.

CHAPTER 16

HAZARD ANALYSIS EXAMPLES

16.1 Introduction

This chapter contains some example case studies intended to help demonstrate and reinforce the concepts presented throughout this book. It should be kept in mind that these examples are slightly contrived and abbreviated for clarity and to prevent requiring extensive time for system understanding.

16.2 Case Study #1 – Hazard Planning

This case study involves hazard planning, and understanding the need for developing a hazard architecture when conducting a HA. This example involves an Ariel Refueling System (ARS) comprised of a Tanker Aircraft (TAC), a Receiver Aircraft (RAC) and the ARS equipment. The ARS subsystems are aboard both the TAC and RAC and synchronize position with each other using GPS and software. A brief system hierarchy diagram is provided in Figure 16.1.

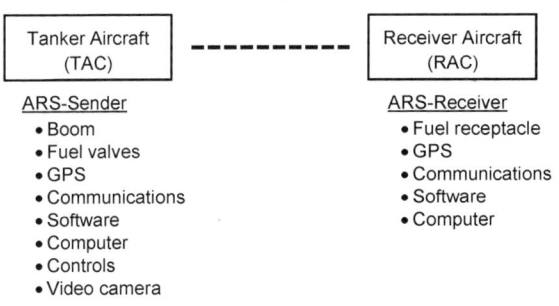

Figure 16.1 – ARS System Diagram

Hazard identification was performed in several iterations, resulting in a HA draft for each iteration. The first draft HA on the ARS system identified the hazards shown in Table 16.1.

Table 16.1 – First Draft Hazards

Loss of fuel containment
Loss of power
Inability of RAC to separate from TAC
Inadvertent separation during refueling
Loss of RAC control due to turbulence from TAC
Collision between TAC and RAC
Fuel spray covers camera
Loss of camera during refueling
Inability of ARS to connect TAC and RAC
EMI interference
FOD
Structural damage
Loss of control
Loss of situational awareness

At first glance these seem like viable hazards, however, after more scrutiny it is apparent that these are not really hazards; they are instead parts of hazards. This list is a confused mix of hazard parts compiled in a shotgun approach, without any critical thinking. These non-hazards demonstrate why it is important to peer review postulated hazards and to also develop a hazard architecture. Hazard architecture helps maintain focus and apply critical thinking to a logical pattern of hazards.

The second HA iteration began the HA process by working at the TLM and TLH levels. This effort resulted in the list of hazards contained in Table 16.2. Note that only one TLM was identified, along with seven TLHs that contribute to that TLM.

Hazard Analysis Primer

Table 16.2 – Second Draft Hazards

TLM	TLH	Hazard
x		ARS causes RAC to crash
	x	Turbulence from TAC causes loss of RAC control
	x	Fuel sprays into RAC engine causing engine fire
	x	ARS boom strikes RAC causing control system damage
	x	ARS unable to provide fuel to RAC, which has < Bingo
	x	ARS boom unable to separate from RAC causing loss of control
	x	Position synchronization errors cause collision between TAC and RAC
	x	FOD causes damage to RAC and loss of RAC control

Table 16.3 is an expansion upon the hazards listed in Table 16.2. In this table, two TLMs were extracted and separated from the original single TLM and added to the list. It appeared that these additional TLMs made more sense than the single TLM. *Bingo* is the minimum amount of fuel available to return to a landing site.

Table 16.3 – Third Draft Hazards

TLM	TLH	Hazard
x		ARS causes RAC to crash
	x	Turbulence from TAC causes loss of RAC control
	x	Fuel sprays into RAC engine causing engine fire
	x	ARS unable to provide fuel to RAC, which has < Bingo
	x	ARS boom unable to separate from RAC causing loss of control
	x	FOD causes damage to RAC and loss of RAC control
x		ARS causes collision between TAC and RAC
	x	Position synchronization errors cause collision
x		ARS causes expensive structural damage to RAC
	x	ARS boom strikes RAC causing damage

Table 17.4 is an expansion upon the hazards listed in Table 16.3. In this table each of the TLHs are expanded into separate hazards contributing to the TLH, with specific causal factors for each unique hazard.

Table 16.4 – Fourth Draft Hazards

TLM	TLH	Hazard
x		ARS causes RAC to crash
	x	Turbulence from TAC causes loss of RAC control - TAC movements cause excessive air turbulence
	x	Fuel sprays into RAC engine causing engine fire - Leak in ARS boom fuel line - ARS boom inadvertently retracts from RAC while fuel flows - Fuel valve fails to close after boom retraction from ARS receptacle
	x	ARS unable to provide fuel to RAC, which has < Bingo - ARS fuel control system fails and RAC < Bingo fuel - ARS inadvertently retracts from RAC and RAC < Bingo fuel
	x	ARS boom unable to separate from RAC causing loss of control - ARS boom control faults prevent separation - ARS receptacle faults prevent separation
	x	FOD causes damage to RAC and loss of RAC control - TFOA TAC strike and damage RAC - TFOA ARS boom strike and damage RAC
x		ARS causes collision between TAC and RAC
	x	Position synchronization errors cause collision - TAC GPS fails and provides false SA position - TAC GPS fails and provides false SA position
x		ARS causes expensive structural damage to RAC
	X	ARS boom strikes RAC causing damage - ARS boom mechanical failures cause swinging and hits RAC - ARS boom operator control system fails - Tanker drogue strikes receptacle and breaks it off
	X	ARS boom structure breaks-up and parts strike RAC - Camera falls off boom

TFOA – things falling off aircraft; FOD – foreign object debris

There are 14 hazards in Figure 16.4, each hazard beginning with a "-" symbol. Given more time and information it may be possible to identify more hazards.

Hopefully, this exercise demonstrates that HA is an iterative process, which requires patience and critical thinking in order to identify and write hazards correctly. Organizing hazards below TLMs and TLHs helps in visualizing and writing hazards.

16.3 Case Study #2 – Hazard Analysis: Coffee Grinder

This case study involves performing a HA on an electric coffee grinder system. Figure 16.2 contains a system diagram for two coffee grinder

Hazard Analysis Primer

system (CGS) configurations. Design A is the proposed design prior to HA and design B is the revised design following HA.

In design A the coffee beans are ground when the unit is plugged into an electrical outlet, beans are placed in the unit and the lid is held down on top of the container unit. The lid pushes the on/off switch down, thereby energizing the unit and the cutting blades. When the lid is removed the switch pops up into the off position, thereby de-energizing the unit.

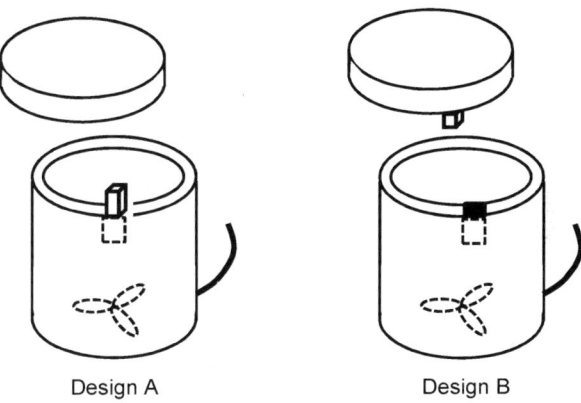

Design A   Design B

Figure 16.2 – Coffee Grinder System (CGS) Diagram

The first step in this HA is to develop some of the HA tools needed to perform a meaningful HA. The basic equipment and functions for this system are identified in Table 16.6. During the conceptual design stage, this is the typical level of information that is available. From this basic design information a very credible list of hazards can easily be generated. Keep in mind that this is a contrived example without any information or knowledge relating to an actual real-life design.

Table 16.6 – HA Information for CGS

Hazard Sources	Functions	Hierarchy Table
Electricity	CGS On	Lid
High speed cutting blades	CGS Off	Container
Exposed ON/OFF button	Grind coffee beans	Cutting blades
	Contains coffee beans	On/Off button
		Power cord

Table 16.5 is a simple SMM for this system. It conveys the major TLMs for the system and their associated hazards.

Table 16.5 – SMM Model

TLM	Hazards for TLM
Fire	Haz-1
Electrocution	Haz-6
Cut Fingers	Haz-2, Haz-3
Disintegration	Haz-4, Haz-5

Hazard-1:
Electrical short in internal wire connections, unit overheats and catches on fire.

- HS – electrical power
- IM – wire short caused by debris, improper connection, etc.
- TTO – house damaged by fire
- HA Clue – HS of electricity
- Mitigation – inspection; non-flammable components

Hazard-2:
Switch failure in ON function mode, blade inadvertently activated while person has finger inside unit; unit is connected to power outlet.

- HS – high speed cutting blades; ON/OFF function
- IM – internal switch failure in the ON mode
- TTO – user injured by cutting blades
- HA Clue – HS of high speed cutting blades
- Mitigation – high quality and reliability parts

Hazard-3:
User accidentally touches exposed button while unit is powered, motor is energized and blades cut off finger of user; unit is connected to power outlet.

- HS – high speed cutting blades
- IM – user touches exposed switch
- TTO – user injury
- HA Clue – high speed cutting blades
- Mitigation – use recessed switch design B

Hazard-4:
Spinning blade comes loose from unit while in use, breaks through plastic housing body and injures user.

- HS – high speed cutting blades
- IM – failure in blade unit housing
- TTO – user injury
- HA Clue – high speed cutting blades
- Mitigation –strong thick plastic housing; high quality parts

Hazard-5:
Spinning blade comes loose from unit while in use, breaks plastic housing and plastic shards injures user.

- HS – high speed cutting blades
- IM – failure in blade unit housing, plastic strength is inadequate
- TTO – user injury
- HA Clue – high speed cutting blades
- Mitigation –strong thick plastic housing; high quality parts

Hazard-6:
Plastic/rubber covering over external wire becomes exposed wire, user touches exposed area receiving electrical shock resulting in electrocution.

- HS – electrical power
- IM – cut in wire covering
- TTO – user electrocution
- HA Clue – HS of electricity
- Mitigation – inspection prior to sale; warning note

As identified by the HA, the main safety problem with design A is that the ON/OFF button is always exposed and inadvertently pressing it without the cover inadvertently energizes the unit and causes the blades to spin. In this scenario there is no protection from the cutting blades. A design safety recommendation is to make the ON/OFF switch recessed, with activator in lid, which is shown in the design B diagram. This requires lid to be closed before blades can operate. The hazard severity remains the same, but the hazard probability of occurrence is reduced. A switch failure could still cause a hazard, but the probability can be reduced by using a high quality switch mechanism.

16.4 Case Study #3 – Hazard Analysis: Aircraft Electrical Power

This case study involves performing a HA on an aircraft electrical power system. Figure 16.3 contains a simplified system diagram of this hypothetical system. The aircraft has two jet engines, each of which powers two electrical generators via bleed air from the engines. A <u>minimum of two</u> generators are required for aircraft electrical power. The system starts with generators G1 and G3 in the operating mode, with G3 and G4 acting as offline backup generators. When monitor M1 detects loss of electrical power from a generator, the computer turns on generator G3 and then G4 as necessary. If G1 fails, the system switches to G2 first, then G4 if G2 is failed. If G3 fails, the system switches to G4 first, then G2 if G4 is failed. Each generator also has internal fault monitoring data which it sends back to the computer, so that the computer can turn on the necessary backup generators. Please keep in mind that this system is relevant but a little over-simplified in order to keep the amount of system detail to a minimum, thereby allowing the focus to be on the HA process.

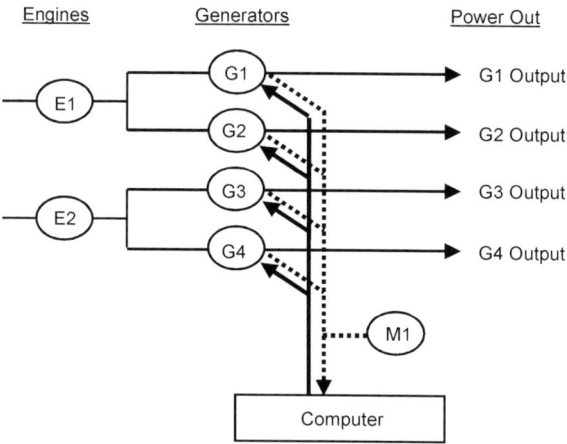

Figure 16.3 – System Functional Diagram

The basic equipment and functions for this system are identified in Table 16.7. During the conceptual design stage, this is the typical level of information that is available. From this basic design information a very credible list of hazards can easily be generated. Keep in mind that this is a contrived example without any information or knowledge relating to an actual real-life design.

Hazard Analysis Primer

Table 16.7 – HA Information for Aircraft Power System

Hazard Sources	Functions	Hierarchy Table
Electrical power	Provide electrical power	Jet engines (2)
High speed turbine blades	Monitor generators	Electrical generators (4)
Jet engines	Switch generators	Computer
Monitor/switching function		Software
		Monitor
		Wiring

Table 16.8 is a simple SMM for this system. It conveys the major TLMs for the system and their associated hazards. The primary undesired TLM for this system is "insufficient electrical power resulting from loss of power from 3 of 4 generators". This means looking at all combinations of three generators failing to provide output to the system. Other TLMs are shown, but have not been developed for this short example.

Table 16.8 – SMM Model

TLM	Hazards for TLM
Loss of Electrical Power	Haz-1 through Haz-9
Fire	Not developed
Loss of all Engines	Not developed

Hazard-1:
Three out of four generators fail to operate, resulting in inadequate aircraft electrical power for flight controls, resulting in aircraft crash. [Note – risk probability calculation will require evaluation of all combinations of 3 out of 4.]

Hazard-2:
Both aircraft engines fail to operate, resulting in loss of all four generators, resulting in inadequate aircraft electrical power for flight controls (and flight), resulting in aircraft crash.

Hazard-3:
Failure of switching unit combined with failure of two Generators, resulting in inadequate aircraft electrical power for flight controls, resulting in aircraft crash. [Note – a risk probability calculation will require evaluation of all combinations.]

Hazard-4:
Failure of Monitor to detect failed Generators combined with failure of two Generators, resulting in inadequate aircraft electrical power for flight controls, resulting in aircraft crash. [Note – risk probability calculation will require evaluation of all combinations.]

Hazard-5:
Wiring from three Generators fails open, resulting in inadequate aircraft electrical power for flight controls, resulting in aircraft crash.

Hazard-6:
Computer software error shuts down power output from all four generators, resulting in inadequate aircraft electrical power for flight controls, resulting in aircraft crash.

Hazard-7:
Computer hardware fault shuts down power output from all four generators, resulting in inadequate aircraft electrical power for flight controls, resulting in aircraft crash.

Hazard-8:
Common-cause failure causes failure of all four generators, resulting in inadequate aircraft electrical power for flight controls, resulting in aircraft crash.

Hazard-9:
Common-cause failure causes failure of both engines, resulting in inadequate aircraft electrical power for flight controls, resulting in aircraft crash.

16.5 Summary

This chapter has covered some example systems and a brief HA for each. Although the examples are brief, one key aspect of HA that was repeatedly demonstrated was the importance of a SMM in the hazard recognition process. For more detailed HA examples refer to reference 1 in Appendix A.

CHAPTER 17

MISHAP INVESTIGATION MODELS

17.1 Introduction

Analyzing a mishap (accident) after it has happened is a difficult process. It seems there is much confusion over what comprises a mishap and how to identify the root-cause factors of a mishap after one has occurred. This confusion may help explain why hazards are misunderstood. This chapter focuses on several models that have been developed to determine the root-causes of mishaps that have occurred. These are essentially *a posteriori* mishap models (after the fact), whereas HA is a priori (before the fact).

17.2 Investigation Models

There are many theories on how to identify the causal factors to a mishap after it has occurred. Benner's observations suggest that there are at least five distinct perceptions of the nature of mishap phenomenon[11]. Each perception results in an accompanying body of assumptions, principles and rules of procedure. These perceptions most likely influence an investigators approach and findings. These models are supported by independent research of Chris Johnson[12].

One thing is evident from the various mishap (accident) models; there is confusion and disagreement on how to represent a mishap. It would seem that a mishap model used by investigators would help define a hazard, yet none of the mishap models presented seem to clearly address hazards. As seen by these mishap models, there is confusion in the mishap

---

[11] Ludwiq Benner. Jr., Five Accident Perceptions and Their Implications for Accident Investigators, Journal of System Safety, Sept-Oct 2009.

[12] Dr. Chris Johnson, A Handbook of Incident and Accident Reporting, Glasgow University Press, ISBN 0-85261-784-4, 2003.

investigation field for how to recognize the causal factors of a mishap. It appears that even investigators are confused about what constitutes a hazard.

There are many mishap investigation techniques that have been proposed and much literature on the subject. Table 17.1 lists the five investigation techniques suggested by Benner.

Table 17.1 – Mishap Investigation Models

Model Name	Description
Single Event	This approach treats the mishap as the result of a single event. The major focus is always on the single event. For example, *human error* is quite often found to be the single event cause. Quite often there are additional factors involved which this approach overlooks.
Chain of Events	This approach treats an accident as a chain of sequential events. It is like the domino effect. This approach is problematic if the causes are forced into a sequence, when all of the causes may not be sequential.
Multi-Linear Events Sequence	This method considers accident events as acting in parallel rather than in series. This approach typically uses event sequence diagrams to show how timing plays into the event chain.
Determinant Variable	This approach applies the statistical analysis of data to reveal the most likely variable (causal factor) causing the accident. This approach is data dependent and may overlook causal factors.
Logic Tree	This approach treats an accident as a converging chain of events. This means the accident is the result of a set of events that may be a combination of sequential and non-sequential in nature. The events can be combined together in a logic tree, such as a FTA.

17.3 Summary

The logic tree, or fault tree, approach seems like the most powerful technique. It looks at all aspects of the mishap and combines the causes into their logical series-parallel set of casual events. The fault tree approach shows that an undesired event (i.e., hazard) has a specific undesired outcome and can be comprised of many different causal factors, combined together in chains of causal factor interrelationships.

One of the questions a mishap investigation should answer is "was this an unidentified or unmitigated hazard?"

CHAPTER 18

SUMMARY

This book has presented and discussed the HA process and many topics associated with hazards and HA. The following are some key points that summarize the HA process:

1) Mishaps are not acts of God, they are acts of carelessness. These acts can be recognized as hazards within a system design, and they can be eliminated or controlled through the HA process.

2) HA is necessary; it must be applied in order to design and develop safe systems. If you do not know what is hazardous in your system, you cannot fix it. Potential mishaps are foreseen via HA.

3) There is a direct link between hazards and mishaps. Potential mishaps are defined by hazards. A mishap cannot occur unless a hazard exists. Hazard magnitude (or amount of danger) is a function of hazard risk.

4) A hazard is comprised of three required components: HS, IM and TTO. These three elements form a Hazard Triangle. If any side can be eliminated (via design), the hazard is eliminated; if a side is mitigated, the hazard risk is then mitigated.

5) When a hazard is mitigated the hazard is not eliminated, the risk is merely reduced. This risk is the *residual* risk.

6) In many cases the system design has a need for a hazardous asset, which can spawn many hazards. These hazards cannot be eliminated; they can only be mitigated (controlled in risk).

7) There are many reasons why hazards exist, such as complexity, failures, human errors, design flaws, use of hazardous assets, etc. It's quite possible that all systems can be found to have some type of hazard, even those systems that seem benign. Do not presume your system is hazard free until a HA has been performed.

8) HA is not a trivial process. There are many complexities involved in the HA process. The HA process should not be taken lightly or just given lip service to fulfill a contract line item.

13) Recognizing or identifying a hazard is typically more difficult than eliminating or mitigating the hazard once it's known. Hazards are somewhat concealed and camouflaged within the system design.

9) HA requires expertise in systems, system safety, hazard theory and HA methodologies.

10) HA must be performed slowly and thoughtfully. Planning the process and establishing the hazard-mishap pattern is important. Do not jump in and start identifying hazards without first establishing a SMM defining the system hazard space. HA requires organization.

11) Quite often it is necessary to write and rewrite a hazard description several times; it is difficult to get it right the first time.

12) Hazards should not be considered too insignificant to enter in the HA worksheets just because the probability may be extremely small. All postulated hazards should be recorded regardless of probability or severity; this shows that everything was considered.

13) Complete hazard identification typically requires more than one type of HA, in order to cover the many diverse components, functions and relationships in a system.

14) A hazard record should not be closed until the safety requirements mitigating the hazard have been verified, and the safety requirements have successfully passed testing.

15) Write hazards to ensure they are characterized in complete system context:

- Describe the Hazard Source (HS), Initiating Mechanism (IM) and Target/Threat Outcome (TTO).
- Do not use abbreviations or assume readers will understand program-special lingo and acronyms.
- Describe the hazard scenario in complete context (e.g., "fuel" is not a hazard, but, "fuel leak and an ignition source leading to fire and system loss" is a hazard).
- Make sure the hazard context includes the specific hazard causal factors and the specific hazard-mishap effects

- Write the hazard in a complete sentence.
- Avoid pseudo hazards (e.g., electrocution)

16) When identifying a hazard, the postulated hazard must pass the following tests:
    - Does the hazard fulfill the Hazard Triangle?
    - Can risk be derived from the hazard description?
    - Is the hazard context clear, concise and complete?
    - Are the hazard causal factors ORed?

17) Risk summing is very complex. If several hazards or hazard causal factors are lumped together, then risk visibility is lost.

18) A design requirement, or system safety requirement (SSR), is established to formally eliminate or mitigate a hazard. Some hazards may require more than one design requirement and some requirements may cover more than one hazard.

19) One hazard ... one mishap. If several different factors can independently cause a hazard, then each set of factors is an individual and separate hazard. These are different hazards with a common mishap outcome; related hazards can be grouped together under TLMs to aid in overall hazard space visibility.

20) A causal factor can be a contributor in more than one hazard, however, each hazard would require additional multiple causal factors in order to be unique.

21) When hazard causal factors are combined with an OR combination, the risk for each cause is summed, thereby increasing the overall hazard risk. When the hazard is broken into several hazards, the risk is lower for each hazard, which more accurately reflects the situation. Also, multiple causal factors confuse hazard mitigation. When a single cause is eliminated or reduced, the risk is still somewhat high from the other causal factors. Therefore, it is best, in general, to break a hazard with multiple ORed causal factors into multiple hazards.

22) One set of causal factors ... one hazard. When a hazard appears to have multiple causes that are ORed together, it is usually a situation involving several hazards, each comprised of one of the sets of causal factors. As depicted in Figure 18.1, a hazard has a single causal factor (Case 1) or multiple causal factors that are ANDed together (Case 2). Case 3 seems like a single hazard

because each of the two hazards has the same common TLM outcome; however, it is actually two hazards. There are, however, occasional exceptions to this generic rule.

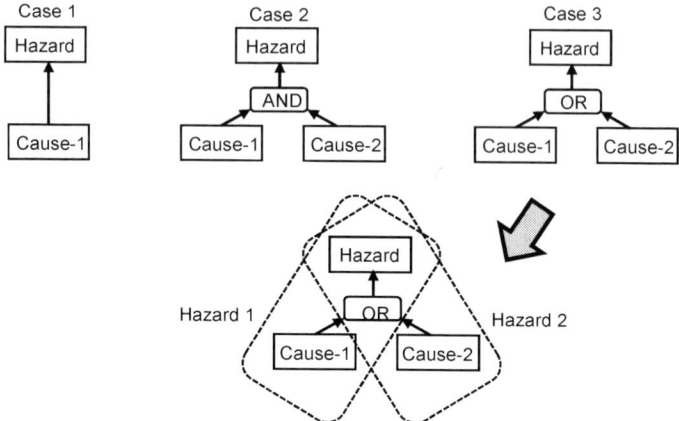

Figure 18.1 – One Hazard vs. Two

APPENDIX A
- MIND MAPPING AND HA -

Mind Mapping (MM) has been around for a number of years for brainstorming, problem solving and organizing thoughts and ideas. MM utilizes a special diagramming technique that drives the thought process for logically identifying all aspects of a topic, including a visual representation of structure and relationships. The mind map diagram is used to represent words, ideas, tasks, or other items linked to and arranged around a central key word or idea. Mind maps are used to generate, visualize, structure, and classify ideas, and as an aid to studying and organizing information for problem solving, decision making and writing. By presenting ideas in a radial, graphical, non-linear manner, mind maps encourage a more effective brainstorming approach. Though the branches of a mind map represent hierarchical tree structures, their radial arrangement disrupts the prioritizing of concepts typically associated with hierarchies presented with more linear visual cues, such as typical outlining structures. Mind maps help you to see both the big picture and the little (detailed) picture, with all the relevant information in between.

Some guidelines for developing a mind map include the following guidelines for creating mind maps:

1. Start in the center with a description of the topic; images can also be included.
2. Select key words and print using upper or lower case letters.
3. Keep the mind map clear by using radial hierarchy to emphasize the branches.
4. Use images, symbols, colors, codes, and dimensions throughout your mind map.
5. Each word/image is best alone and sitting on its own line.
6. The lines should be connected, starting from the central topic.
7. Colors can be used for visual stimulation and to group items.
8. Develop your own personal style of mind mapping.
9. Use emphasis and show associations.

Mind maps can be drawn manually or by hand, or for better quality a computer generated diagram can be applied. There are a number of software packages available for producing mind maps, both free and purchased. Figure A.1 contains an example mind map diagram for the author's background, which could be used to help me write my biography.

It demonstrates the brainstorming and organizational capabilities of MM. This mind map was drawn using one of the free tools available.

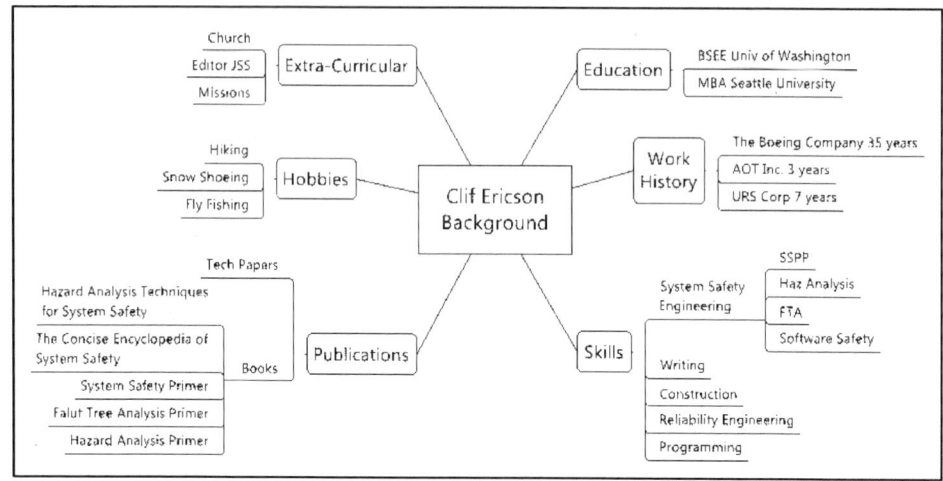

Figure A.1 – Example Mind Map

The brainstorming and organizational capabilities of MM are directly applicable to the HA process. In fact, I believe that MM would help in laying out a SMM that would greatly aid in the identification and description of hazards. To validate this theory I have developed a progression on mind maps diagrams for a HA of an automobile. I chose the automobile system because it is ubiquitous and can be fairly well understood, at a high level, without a set of schematics and drawings.

Figure A.2 is the first cut at a HA of an automobile. In the center is the topic system of the HA. The first layer of branches identifies the TLMs of concern for the system. These TLMs identify the major undesired mishap outcomes for the system; they establish the paths the HA will follow and initiates the organization of hazards to be identified.

Figure A.3 is the second cut at the HA of an automobile. It analyzes each TLM and develops the paths, with content, to the final hazard. At this stage, it may take one or more sub-branches until the final hazard is reached.

Figure A.4 is the third cut at the HA of an automobile. In this mind map two branches are developed to the end until the associated hazards are identified. Two hazards are developed in this example; note that the hazard description contains the HS, IM and TTO elements of hazard.

Hazard Analysis Primer

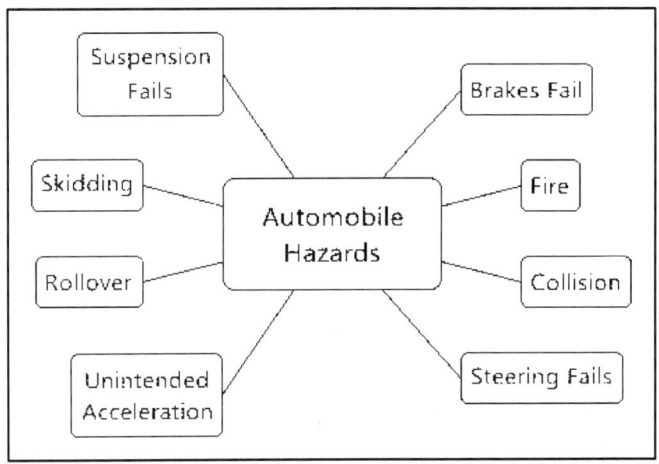

Figure A.2 – First Cut HA Mind Map

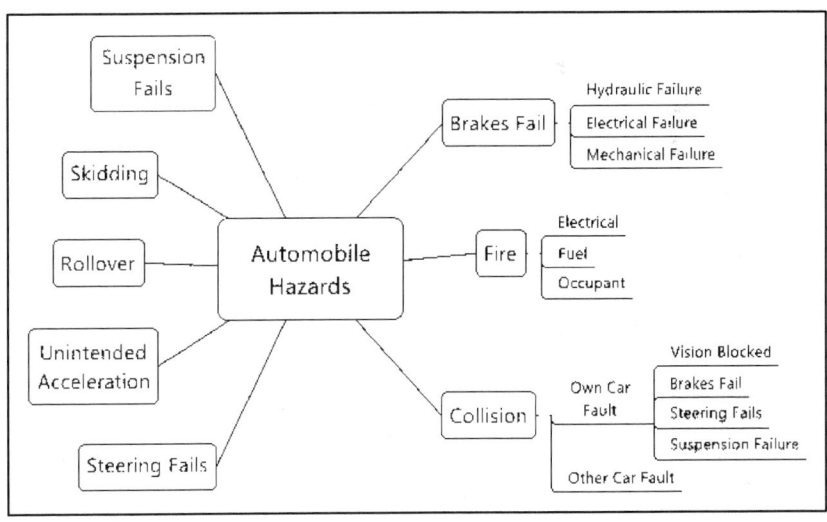

Figure A.3 – Second Cut HA Mind Map

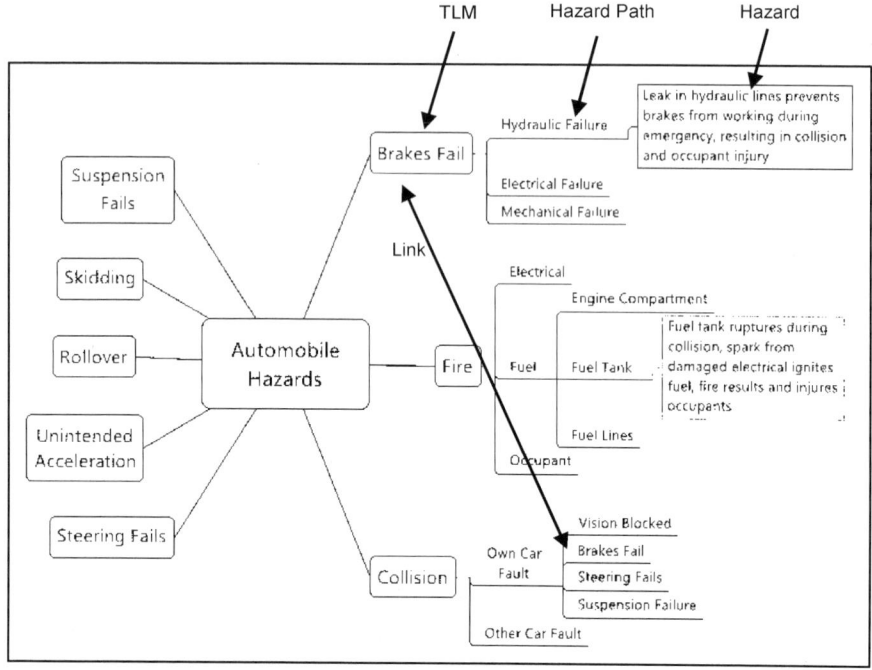

Figure A.4 – Third Cut HA Mind Map

Note that the mind map develops important relationships that must be resolved. For example, the TLM Brake Failure is an important TLM that is of high interest by itself; however, it also could fall under the Collision TLM. As each TLM path is further expanded, the conditions, factors and causal factors to each hazard is gradually developed.

Once the TLM-hazard path layout is established, the actual final hazards can be written on a spreadsheet or HA forms, as they will begin to take up considerable space in the mind map.

As pointed out throughout the book, it is imperative to develop a SMM, such as the mind map, before establishing the final set of hazards. The MM approach helps to HA avoid performing activity without insight.

APPENDIX B
- REFERENCES -

There are no official standards for performing a HA. There are many books and papers on the topic, however, most do not provide adequate detail, except for reference 1, which describes 24 of the most used HA techniques in extensive detail. It elaborates on the steps for each methodology, errors often made, pros and cons, and examples for each technique. The HA techniques described include both primary and secondary methods.

The following books are some of the most relevant and useful resources on HA:

1) Hazard Analysis Techniques for System Safety, C. A. Ericson II, John Wiley and Sons, 2005.

2) System Safety Primer, C. A. Ericson II, printed by CreateSpace, 2011.

3) Fault Tree Analysis Primer, C. A. Ericson II, printed by CreateSpace, 2011.

4) System Safety Engineering and Management, H. E. Roland and B. Moriarity, Wiley and Sons, 1990.

5) System Safety Analysis Handbook, System Safety Society, www.system-safety.org.

6) System Safety: HAZOP and Software HAZOP, F. Redmill, M. Chudleigh and J. Catmur, John Wiley and Sons, 1999.

7) Job Hazard Analysis: A guide for voluntary compliance and beyond, J. Roughton and N. Crutchfield, Butterworth-Heinemann, 2007.

8) Safeware: System Safety and Computers, N. G. Leveson, Addison-Wesley Publishing, 1995.

9) Assurance Technologies Principles and Practices, D. G. Raheja and M. Allocco, John Wiley and Sons, 2006.

10) MIL-STD-882, Standard Practice for System Safety, Version D, 10 February 2000 (Original version: 15 July 1969).

11) ANSI/GEIA-STD-0010-2009, Standard Best Practices for System Safety Program Development and Execution, 12 February 2009.

12) Concise Encyclopedia of System Safety: Definition of Terms and Concepts, C. A. Ericson II, John Wiley and Sons, 2011.

13) Mindmapping: Your Personal Guide to Exploring Creativity and Problem-Solving, Joyce Wycoff, Berkley Trade, 1991.

14) Use Both Sides of Your Brain: New Mind-Mapping Techniques, Tony Buzan, Plume, Third Edition, 1991.

APPENDIX C
- ABOUT THE AUTHOR -

Mr. Ericson has over 45 years of experience in the field of system safety, software design, software safety and Fault Tree Analysis (FTA). He holds a BSEE from the University of Washington and an MBA from Seattle University. Currently he works for the URS Corporation (formerly EG&G Technical Services) in Dahlgren, VA. He provides technical analysis, consulting, oversight and training on system safety and software safety projects. He currently supports NAVAIR system safety on the UCAS and BAMS unmanned aircraft systems, and he is assisting in writing NAVAIR system safety policies and procedures. Prior to joining URS, Mr. Ericson worked at Applied Ordnance Technology (AOT), Inc. of Waldorf, Maryland, where he was a program manager of system and software safety. In this capacity he directed projects in system safety and software safety engineering.

Prior to joining AOT, Mr. Ericson was employed as a Senior Principal Engineer for the Boeing Company for 36 years. At Boeing he worked in the fields of system safety, reliability, software engineering and computer programming. Mr. Ericson has been involved in all aspects of system safety, including hazard analysis, FTA, software safety, safety certification, safety documentation, safety research, new business proposals and safety training. He has worked on a diversity of projects, such as the Minutemar Missile System, SRAM missile system, ALCM missile system, Morgantov People Mover system, 757/767 aircraft, B-1A bomber, AWACS syst Boeing BOECOM system, EPRI solar power system and the A' Technical Integration program.

Mr. Ericson has taught courses on software safety and FTA University of Washington. Mr. Ericson was President of the Syste Society in 2001-2003, and served as Executive Vice President of t' Safety Society, and Co-Chairman of the 16th International Sy Conference. He was the technical program chairman for the 19 International System Safety Conferences. He is the founder Sound chapter (Seattle) of the System Safety Society. In 200 Apollo Award for safety consulting work on the Interna Station, and the Boeing Achievement Award for developing th course. Mr. Ericson won the System Safety Society' Achievement Award in 1998, 1999 and 2004 for outstandin system safety field.

Mr. Ericson has prepared and presented training courses in system safety, software safety and FTA in the U.S., Singapore and Australia and has presented numerous technical papers at safety conferences. Mr. Ericson has published many technical articles on system and software safety and is currently editor of the Journal of System Safety (JSS), a publication of the International System Safety Society. Mr. Ericson is the author of the NAVSEA Weapon System Safety Guidelines Handbook. Other books published by Mr. Ericson include:

1) Hazard Analysis Techniques for System Safety, July 2005, John Wiley, Inc.

2) Concise Encyclopedia of System Safety: Definition of Terms and Concepts, July 2011, John Wiley, Inc.

3) System Safety Primer, September 2011, printed by CreateSpace, Inc.

4) Fault Tree Analysis Primer, December 2011, printed by CreateSpace, Inc.

Mr. Ericson can be reached through his website at www.risk-logic.com.

APPENDIX D
- INDEX -

Barrier Analysis (BA)	98, 105
Bent Pin Analysis (BPA)	98, 104
Common Cause Failure (CCF)	106, 113
Common Cause Failure Analysis (CCFA)	98, 106
Event Tree Analysis (ETA)	98, 104
Exposure Time	57
Failure Data	135
Failure Modes and Effects Analysis (FMEA)	28, 98, 104
Fault Tree Analysis (FTA)	98, 103
Functional Flow Block Diagram	26
Functional Hazard Analysis (FHA)	98, 103
Hazard Analysis	4, 69, 87
Hazard and Operability Analysis (HAZOP)	98, 103
Hazard Causal Factor (HCF)	121
Hazard Risk Index (HRI)	48, 74, 151
Hazard Tracking System (HTS)	75, 94
Hazard Triangle	40, 81
Hazard	31, 44
Health Hazard Assessment (HHA)	98, 103
HS-IM-TTO Hazard Model	14, 31
Interlock Analysis	98, 106
Mind Mapping	179
Operations and Support Hazard Analysis (O&SHA)	98, 102
PLOA Table	29
Preliminary Hazard Analysis (PHA)	98, 101
Preliminary Hazard List (PHL) Analysis	98, 101
Primary Hazard Analysis	100, 141
Probabilistic Risk Assessment (PRA)	98, 107
Probability of Loss of Aircraft (PLOA)	29
Reliability Block Diagram (RBD)	27
Reliability	136
Risk	31, 45
Root Cause Analysis (RCA)	69
Safety Assessment Report (SAR)	98, 107
Safety Requirements/Criteria Analysis (SRCA)	98, 105
Secondary Hazard Analysis	100, 141
Simplified System Diagram	25
S-M-O Hazard Model	14, 31
Sneak Circuit Analysis (SCA)	98, 105
Subsystem Hazard Analysis (SSHA)	98, 102
System Hazard Analysis (SHA)	98, 102
System Hierarchy Table	23

System Lifecycle	21
System Mishap Model (SMM)	64, 179
System Safety	1
System Safety Requirement (SSR)	55, 74, 105, 177
System Views	21
System	18
Tailoring (of HA techniques)	142
Tailoring (of Risk Indices)	150
Test Hazard Analysis	98, 106
Threat Hazard Assessment (THA)	98, 106
Top Level Mishap (TLM)	63
Undesired Event (UE)	2, 84

Made in the USA
Columbia, SC
12 March 2018